Springer Theses

Recognizing Outstanding Ph.D. Research

For further volumes:
http://www.springer.com/series/8790

Aims and Scope

The series "Springer Theses" brings together a selection of the very best PhD theses from around the world and across the physical sciences. Nominated and endorsed by two recognised specialists, each published volume has been selected for its scientific excellence and the high impact of its contents for the pertinent field of research. For greater accessibility to non-specialists, the published versions include an extended introduction, as well as a foreword by the student's supervisor explaining the special relevance of the work for the field. As a whole, the series will provide a valuable resource both for newcomers to the research fields described, and for other scientists seeking detailed background information on special questions. Finally, it provides an accredited documentation of the valuable contributions made by today's younger generation of scientists.

Theses are accepted into the series by invited nomination only and must fulfill all of the following criteria

- They must be written in good English.
- The topic should fall within the confines of Chemistry, Physics, Earth Sciences, Engineering and related interdisciplinary fields such as Materials, Nanoscience, Chemical Engineering, Complex Systems and Biophysics.
- The work reported in the thesis must represent a significant scientific advance.
- If the thesis includes previously published material, permission to reproduce this must be gained from the respective copyright holder.
- They must have been examined and passed during the 12 months prior to nomination.
- Each thesis should include a foreword by the supervisor outlining the significance of its content.
- The theses should have a clearly defined structure including an introduction accessible to scientists not expert in that particular field.

Matthew Joseph Mottram

A Search for Ultra-High Energy Neutrinos and Cosmic-Rays with ANITA-2

Doctoral Thesis accepted by
University College London, UK

Springer

Author
Matthew Joseph Mottram
University College London
London
UK

Supervisor
Dr. Ryan Nichol
University College London
London
UK

ISSN 2190-5053 ISSN 2190-5061 (electronic)
ISBN 978-3-642-30031-8 ISBN 978-3-642-30032-5 (eBook)
DOI 10.1007/978-3-642-30032-5
Springer Heidelberg New York Dordrecht London

Library of Congress Control Number: 2012940725

Printed on acid-free paper

Springer is part of Springer Science+Business Media (www.springer.com)

Supervisor's Foreword

Almost exactly 100 years ago, Victor Hess undertook a series of pioneering, not to mention dangerous, balloon flights and discovered the mysterious cosmic rays that are continuously bombarding the Earth's atmosphere. After a century of study, the origin and nature of the highest energy cosmic rays is still a mystery to scientists. One missing piece in the cosmic ray puzzle is the detection of cosmogenic neutrinos, which originate from the interactions of the highest energy cosmic rays with photons of the cosmic microwave background. The ANITA experiment was designed to dangle from a long-duration balloon, almost 40 km above Antarctica, and detect the radio signals produced when these neutrinos interact in the ice below.

The Antarctica Impulsive Transient Antenna (ANITA) experiment is at the frontier of three areas of scientific endeavour: the search for ultra-high energy neutrinos of astrophysical origin; the characterisation of ultra-high energy cosmic rays and the development of experimental techniques for the radio detection of high energy particles.

In this thesis, Matthew Mottram describes the preparation for the 2008 flight of the ANITA instrument and the results of the searches for signals from ultra-high energy neutrinos and cosmic rays in the data. By the time you finish reading this work, you will not only know the outcomes of these analyses, but should also have some feeling for the level of painstakingly detailed work required to make a modern astroparticle physics experiment work. Hopefully, you will also realise that working at the frontiers of science is not only challenging but is also hugely enjoyable and rewarding. After all, how many people get to fly balloons around Antarctica looking for particles from outer space?

April 2012 Dr. Ryan Nichol

Acknowledgments

First and foremost I would like to thank my supervisor, Ryan Nichol, who has been a constant source of advice and ideas throughout my time at UCL. I would also like to thank all the ANITA collaborators, particularly Abigail Vieregg, Stephen Hoover, and Andres Romero-Wolf for their advice regarding my analysis and Amy Connolly for helping with the simulation code. Further thanks to David Saltzberg and Peter Gorham for all the help they provided throughout my PhD. Thanks to Brian Hill for the image from the ANITA-2 landing site. Brian Mercurio carried out the work on ANITA-2's directional exposure that provided the basis for the point source limits in Chap. 8 while Dave Besson provided data used for the RICE experiment limits, thanks to you both. Thanks also to the Columbia Scientific Balloon Facility guys, without whom ANITA could not have got off the ground.

Thanks to the all the members of the UCL HEP group. From seemingly endless coffee breaks through to worryingly forgotten nights in the Jeremy Bentham, my liver and I have thoroughly enjoyed our time in the group.

Thanks to my parents for their unwavering support. You always tell me how impressed you are that I should have chosen to study for a PhD in physics, but none of it would have been possible without your encouragement and belief. Finally, thank you Jo, I cannot imagine what the last 4 years would have been like without you there to calm nerves and provide motivation.

Contents

Acronyms

ADC	Analogue to digital convertor
AGN	Active galactic nuclei/nucleus
ANITA	The ANtarctic impulsive transient Antenna
CSBF	Columbia scientific balloon facility
CW	Continuous wave
EAS	Extensive air shower
GPS	Global positioning system
GRB	Gamma-ray burst
GZK	Griesen, Zatsepin and Kuzmin
HPOL	Horizontally polarised
LNA	Low-noise amplifier
LoS	Line of sight
PID	Proportional integral differential
PPS	Pulse per second
RF	Radio frequency
RFCM	Radio frequency conditioning module
RMS	Root mean square
SCA	Switched capacitor array
SHORT	SURF high occupancy RF trigger
SNR	Signal to noise
SURF	Sampling unit for radio frequencies
TURF	Trigger unit for radio frequencies
UHE	Ultra-high energy
UHECR	Ultra-high energy cosmic-ray
VPOL	Vertically polarised

Chapter 1
Introduction

The high energy Universe remains mysterious. cosmic-rays have been observed at ultra-high energies (UHE, $E > 10^{18}$ eV) for half a century, yet the mechanisms of acceleration of such particles are poorly understood. At the highest energies traditional astrophysical messengers (cosmic-rays and photons) suffer horizon effects that limit their use. Protons and heavier nuclei interact with cosmic microwave background radiation respectively via photo-pion production and photo-disintegration, while photons pair-produce with the far infra-red background and cosmic microwave background.

Neutrinos provide a potential solution to this UHE observation horizon. As they only interact weakly, UHE neutrinos are expected to traverse cosmic distances virtually uninhibited. Unlike cosmic-rays, whose paths will be affected by the presence of Galactic and extra-galactic magnetic fields, neutrinos suffer no such deviations in transit. Observation of UHE neutrinos therefore provides a possible window on the highest energy processes in the Universe. Meanwhile, the very mechanism that is expected to result in the UHE cosmic-ray horizon will also give rise to a neutrino flux, through the pion decay chain. Given the observational evidence for UHE cosmic-rays, an expected (so called 'guaranteed') UHE neutrino flux exists.

Due to the neutrino's small cross section and low flux in the UHE regime, the collection of a statistically significant sample of UHE neutrinos is a considerable challenge. Novel detection techniques and concepts have been developed and tested over the last two decades but, as yet, the observation of UHE neutrinos remains an unachieved goal of both astrophysics and particle physics. The ANtarctic Impulsive Transient Antenna (ANITA) hopes to detect UHE neutrinos through coherent radio Cherenkov emission from particle showers in dense media. This coherent Cherenkov emission process, known as Askaryan radiation, has been experimentally confirmed through a number of accelerator experiments. Supported by a giant helium balloon, ANITA observes the radio transmissive Antarctic ice sheet at an altitude of $\sim$35 km. ANITA is the most sensitive UHE neutrino experiment at $E > 10^{19}$ eV.

Two science flights of the ANITA experiment have taken place. Analysis of data from the first flight, hereafter ANITA-1, observed no evidence of neutrino emission.

M. J. Mottram, *A Search for Ultra-High Energy Neutrinos and Cosmic-Rays with ANITA-2*, Springer Theses, DOI: 10.1007/978-3-642-30032-5_1,

Analysis of the second flight, hereafter ANITA-2, is ongoing. The first published results from ANITA-2 set the most stringent limit on the UHE neutrino flux in the energy interval $10^{19} < E_{\nu} < 10^{21}$ eV to date.

The work described in this thesis includes an analysis of the ANITA-2 data, including searches for both UHE neutrinos and UHE cosmic-rays. An introduction to UHE neutrino and cosmic-ray physics, along with detection techniques, is given in Chap. 3. A detailed description of the ANITA-2 experiment is given in Chap. 4. Chapter 5 covers an ANITA-2 instrument simulation developed for use with the analysis code that is described in Chap. 6. Finally, Chaps. 7 and 8 describe further analysis of candidate cosmic-ray and neutrino events respectively, with Chap. 8 providing constraints on the diffuse UHE neutrino flux, as well as constraints on the flux from selected sources.

Chapter 2
Particle Physics and the Neutrino

2.1 Introduction to Particle Physics

The Standard Model of particle physics has been incredibly successful at describing fundamental particles and their interactions, for a comprehensive introduction and overview see e.g. [1–3]. The Standard Model is an $SU(3) \otimes SU(2) \otimes U(1)$ gauge theory that consists of fermions (quarks and leptons) and fundamental forces (electromagnetic, weak nuclear and strong nuclear) that are mediated by force carrying particles, bosons.

2.1.1 Quarks

There are six different types (or flavours) of quarks which are divided into three generations:

$$\begin{pmatrix} u \\ d \end{pmatrix}, \begin{pmatrix} c \\ s \end{pmatrix}, \begin{pmatrix} t \\ b \end{pmatrix} \tag{2.1.1}$$

where u, c, t have charge $+\frac{2}{3}$ and d, s, b have charge $-\frac{1}{3}$. The generations are grouped by mass, with u, d being the least and t, b the most massive. Quarks are fermions and have $\frac{1}{2}$-integer spin. Quarks also possess one of three colour charges, a property of quantum chromodynamics that relates to the strong-nuclear force interactions.

2.1.2 Leptons

As with quarks, leptons are divided into three generations, which are defined by their flavour:

M. J. Mottram, *A Search for Ultra-High Energy Neutrinos and Cosmic-Rays with ANITA-2*, Springer Theses, DOI: 10.1007/978-3-642-30032-5_2,

$$\begin{pmatrix} e \\ \nu_e \end{pmatrix}, \begin{pmatrix} \mu \\ \nu_\mu \end{pmatrix}, \begin{pmatrix} \tau \\ \nu_\tau \end{pmatrix} \tag{2.1.2}$$

The upper particle in each flavour set is a lepton with charge -1, the lower particle, ν_i, is a chargeless neutrino of the corresponding flavour. The charged leptons above are ordered by mass, with e being the least and τ the most massive. For a description of neutrino masses, see Sect. 2.2. Leptons are fermions, with $\frac{1}{2}$-integer spin. Leptons are colourless and as such do not interact via the strong force.

2.1.3 Bosons

The four fundamental forces are mediated via force carriers which have integer spin, called bosons. The electromagnetic force is mediated via the massless photon, γ. The electromagnetic force will only couple to those particles that carry charge. The strong nuclear force is mediated via gluons, g, and couples to particles with colour charge. The strength of the strong force increases with separation, resulting in confinement of quarks in colourless groupings known as hadrons. The weak nuclear force, transmitted via three bosons $W^{\pm}$ and Z, couples to particles with weak isospin. All quarks and leptons carry weak isospin. Although the fundamental strength of the weak force is of the same order as the electromagnetic force, the massive nature of the $W^{\pm}$ and Z, 80.4 and 91.2 GeV respectively, makes the force appear weak and short ranged. Gravity is the weakest force and, although it is not included in the Standard Model, is thought to be mediated by a spin 2 boson known as a graviton.

Quarks, therefore, couple to all fundamental forces. Charged leptons do not feel the strong force. Neutrinos in the Standard Model, meanwhile, feel only the weak force.

2.2 The Neutrino

The existence of the neutrino was first predicted by Wolfgang Pauli in 1930 to explain the apparent discrepancy between initial and final energy and momenta in beta decays. The particle, initially dubbed the neutron by Pauli, was named *neutrino* (little neutral one) by Enrico Fermi in 1934, after the discovery and naming of the heavier particle we know as the neutron.

Neutrinos are known to exist in three flavour states (ν_e, ν_μ and ν_τ) as partners to the three, heavier, charged leptons (e, μ and τ). Much of the neutrino's nature is yet to be established; investigation into the neutrino is seen as one of the most promising avenues for beyond Standard Model physics.

As neutrinos interact only weakly, they are only able to exchange one of the three weak force bosons, $W^{\pm}$ and Z^0, with these interactions known as

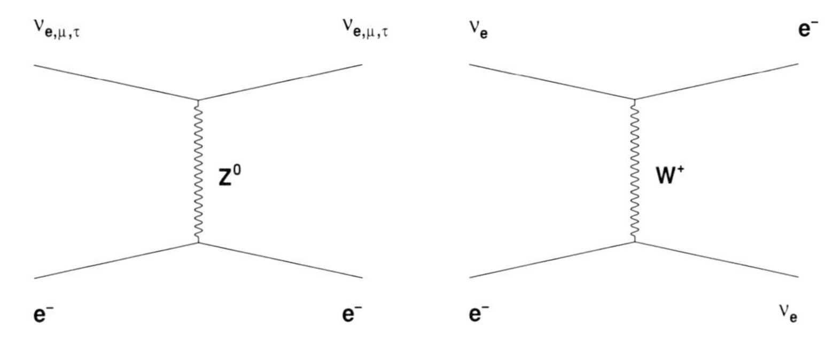

Fig. 2.1 Feynman diagrams of neutral current (*left*) and charged current (*right*) neutrino interactions with electrons

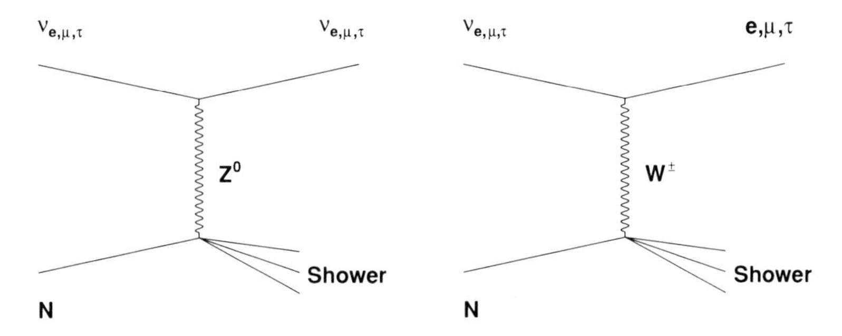

Fig. 2.2 Feynman diagrams of neutral current (*left*) and charged current (*right*) neutrino interactions with nucleons

charged-current (CC) and neutral current (NC) respectively. It is these interactions that enable experiments to observe neutrinos, with observations of secondary leptons or through the search for hadronic recoil. Feynman diagrams of neutrino interactions are given in Figs. 2.1 and 2.2.

2.2.1 Neutrino Oscillations

The three neutrino flavour states are themselves superpositions of three mass states, m_1, m_2 and m_3, with the relation between the flavour (α) and mass (i) states given by:

$$|\nu_\alpha\rangle = \sum_i U_{\alpha i} |\nu_i\rangle \tag{2.2.1}$$

Here, U is the PMNS mixing matrix (named for Pontecorvo, Maki, Nakagawa and Sakata), which defines neutrino mixing and oscillations:

$$U = \begin{pmatrix} c_{12}c_{13} & s_{12}c_{13} & s_{13}e^{-i\delta} \\ -s_{12}c_{23} - c_{12}s_{23} & c_{12}c_{13} - s_{12}s_{23}s_{13}e^{i\delta} & s_{23}c_{13} \\ s_{12}s_{23} - c_{12}c_{23}s_{13}e^{i\delta} & c_{12}s_{13} - s_{12}c_{23}s_{13}e^{i\delta} & c_{23}c_{13} \end{pmatrix} \begin{pmatrix} 1 & 0 & 0 \\ 0 & e^{i\alpha} & 0 \\ 0 & 0 & e^{i\beta} \end{pmatrix} \tag{2.2.2}$$

where $s_{ij} = \sin(\theta_{ij})$, $c_{ij} = \cos(\theta_{ij})$ and $e^{i\delta}$ is a CP violating term with phase δ. The terms α and β are Majorana CP violating phases. The matrix can be factorised as follows:

$$U = \begin{pmatrix} 1 & 0 & 0 \\ 0 & c_{23} & s_{23} \\ 0 & -s_{23} & c_{23} \end{pmatrix} \begin{pmatrix} c_{13} & 0 & s_{23}e^{-i\delta} \\ 0 & 1 & 0 \\ -s_{13}e^{i\delta} & 0 & c_{13} \end{pmatrix} \begin{pmatrix} c_{12} & s_{12} & 0 \\ -s_{12} & c_{12} & 0 \\ 0 & 0 & 1 \end{pmatrix} \begin{pmatrix} 1 & 0 & 0 \\ 0 & e^{i\alpha} & 0 \\ 0 & 0 & e^{i\beta} \end{pmatrix} \tag{2.2.3}$$

Equations 2.2.2 and 2.2.3 demonstrate that the three states of neutrinos will mix with one another. After creation in a flavour state, the neutrino will then propagate as a superposition of the mass eigenstates. Any subsequent interaction will then take place as a flavour state, with the interacting flavour dependent on the mixing parameters. This gives rise to the concept of neutrino flavour oscillation.

It is useful to consider a two flavour system to see how this mass propagation effects the flavour state measured by an observer. In this simplified case, a mixing matrix would be given by:

$$\begin{pmatrix} \nu_\alpha \\ \nu_\beta \end{pmatrix} = \begin{pmatrix} \cos(\theta) & \sin(\theta) \\ -\sin(\theta) & \cos(\theta) \end{pmatrix} \begin{pmatrix} \nu_1 \\ \nu_2 \end{pmatrix} \tag{2.2.4}$$

A neutrino created in weak eigenstate ν_α is therefore a combination of the two mass states ν_1 and ν_2, as defined by the mixing angle θ:

$$|\nu_\alpha\rangle = \cos(\theta)|\nu_1\rangle + \sin(\theta)|\nu_2\rangle \tag{2.2.5}$$

After propagating distance L, this neutrino will be in the following state:

$$|\nu_{x=L}\rangle = \cos(\theta)e^{iE_1t}|\nu_1\rangle + \sin(\theta)e^{iE_2t}|\nu_2\rangle \tag{2.2.6}$$

where E_1, E_2 are the energies of the two mass eigenstates and e^{iE_1t}, e^{iE_2t} are their propagation as a function of time. The probability of an oscillation into the other flavour eigenstate when measured is thus:

$$\begin{aligned} P_{\nu_\alpha \to \nu_\beta} &= \left|\langle \nu_\beta|\nu_\alpha(t)\rangle\right|^2 \\ &= \left|(-\sin(\theta)\langle\nu_1| + \cos(\theta)\langle|\nu_2|)\left(\cos(\theta)e^{iE_1t}\nu_1\rangle + \sin(\theta)e^{iE_2t}|\nu_2\rangle\right)\right|^2 \\ &= \left|\left(-\sin(\theta)\cos(\theta)e^{iE_1t} + \cos(\theta)\sin(\theta)e^{iE_2t}\right)\right|^2 \\ &= \left|\cos(\theta)\sin(\theta)\left(e^{iE_1t} - e^{iE_2t}\right)\right|^2 \\ &= \sin^2(2\theta)\sin^2\left(\frac{E_2 - E_1}{2}t\right) \end{aligned} \tag{2.2.7}$$

Assuming $E_i \gg m_i$, $E_i = \sqrt{m_i^2 + p^2}$, the probability of an oscillation becomes:

$$P_{\nu_\alpha \to \nu_\beta} = \sin^2(2\theta)\sin^2\left(\frac{1.27\Delta(m_{12}^2)L}{E}\right) \quad (2.2.8)$$

where $\Delta\left(m_{12}^2\right)$ is the difference in mass between the two mass eigenstates in eV, L is the distance travelled in km and E is the neutrino energy in GeV.

2.2.2 *Measurement of Mixing Parameters*

It was the observation of a flux deficit of Solar ν_e, first observed by the Homestake Mine experiments in the 1960s and 1970s [4], that led to the realisation of neutrino flavour oscillation. Further experiments such as the Sudbury Neutrino Observatory (SNO) [5] were designed to measure the solar neutrino flux. This led to the determination of the so called 'Solar' neutrino oscillation parameters ($\nu_e \to \nu_{\mu,\tau}$ for θ_{12} and $\Delta\left(m_{12}^2\right)$).

Other experiments, such as Kamiokande [6] and Super-Kamiokande [7], use cosmic-ray-induced $K^\pm$ and $\pi^\pm$ decay chains (Eq. 2.2.9) to measure the 'atmospheric' oscillation parameters ($\nu_\mu \to \nu_\tau$ for θ_{23} and $\Delta\left(m_{23}^2\right)$):

$$\pi^+\left(\pi^-\right) \to \mu^+\nu_\mu\left(\mu^-\bar{\nu}_\mu\right) \quad (2.2.9)$$

$K^\pm$ will decay to produce the same end state, but may also decay to produce final states containing $\pi^0\mu^\pm\overset{(-)}{\nu}_\mu$ or $\pi^0e^\pm\overset{(-)}{\nu}_e$. Any $\mu^\pm$ produced will decay (although while the meson decay will occur in the atmosphere, the resultant μ may penetrate into the Earth).

Equation 2.2.8 demonstrates that, for two flavour neutrino oscillations, there will be an optimal L/E value at which mixing will be maximal. In the last decade a number of experiments, e.g [8–10], have taken advantage of this, using neutrinos produced at accelerators to make precision measurements of both atmospheric and solar neutrino oscillation parameters, as well as place limits on the remaining oscillation parameters, θ_{13} and $\Delta\left(m_{13}^2\right)$ (see Table 2.1).

2.2.3 *Neutrino Mass*

Neutrinos are known to have finite mass, otherwise flavour oscillation would not be possible. While neutrino oscillation measurements provide the absolute difference in mass between the neutrino mass eigenstates, the ordering of these masses is not known. This is known as the neutrino mass hierarchy problem.

Table 2.1 Neutrino oscillation parameters, taken from [11] unless stated otherwise. θ_{ij} and $\Delta\left(m_{ij}^2\right)$ are the mixing angles and the squared mass splittings respectively

Parameter	Value
$\sin^2(2\theta_{12})$	$0.861^{+0.026}_{-0.022}$
$\Delta\left(m_{12}^2\right)$	$(7.59 \pm 0.21) \times 10^{-5}\,\text{eV}^2$
$\sin^2(2\theta_{23})$	>0.92
$\Delta\left(m_{23}^2\right)$	$(2.32^{+0.12}_{-0.08}) \times 10^{-3}\,\text{eV}^2$ [12]
$\sin^2(2\theta_{13})$	<0.15

Table 2.2 Constraints on the neutrino mass

Parameter	Mass limit	Source
$m_{i,heaviest}$	$>0.05\,\text{eV}$	Oscillation (Δm_{23}^2) [11]
Σm_i	$<2\,\text{eV}$	Cosmology [11]
m_β	$<2.0\,\text{eV}$	β-decay [14]

Limits on the mass eigenstates have been obtained through neutrino oscillation experiments and cosmological[1] measurements. Experimental measurements of the energy spectrum endpoint of β decay provides a further constraint on the electron neutrino mass, m_{ν_e}. These values are summarised in Table 2.2.

As the neutrino has mass, it is possible for the neutrino to be its own anti-particle, known as a Majorana particle. In this scenario ν are purely left-handed particles and $\bar{\nu}$ are purely right-handed (i.e. ν viewed in a frame in which its helicity is flipped). It is possible to determine whether the neutrino is a Majorana particle through measurements of double β-decay, specifically searching for neutrinoless double β-decay. Through these experiments, combined with further cosmological measurements and single β-decay measurements, it is hoped that constraints on neutrino mass, neutrino mass hierarchy and the Majorana/Dirac nature of the neutrino, will be found.

References

1. F. Halzen, A.D. Martin, *Quarks and Leptons* (Wiley, New York, 1985)
2. W.N. Cottingham, D.A. Greenwood, *An Introduction to the Standard Model of Particle Physics* (Cambridge University Press, Cambridge, 1999)
3. I.J.R. Aitchison, A.J.G. Hey, *Gauge Theories in Particle Physics* (Taylor & Francis, New York, 1989)
4. B.T. Cleveland et al., Astrophys. J. **496**, 505 (1998)
5. The SNO Collaboration, S.N. Ahmed et al., Phys. Rev. Lett. **92**, 181301 (2004), [nucl-ex/0309004]

[1] Galaxy surveys and anisotropy analysis in the CMB both place limits of $\Sigma m_i < 2\,\text{eV}$, combinations of multiple cosmological probes place a limit of $\Sigma m_i < 0.17\,\text{eV}$ at the 95 % CL [13].

6. The Kamiokande Collaboration, S. Hatakeyama et al., Phys. Rev. Lett. **81**, 2016 (1998), [hep-ex/9806038]
7. The Super-Kamiokande Collaboration, Y. Ashie et al., Phys. Rev. D **71**, 112005 (2005)
8. The K2K Collaboration, M.H. Ahn et al., Phys. Rev. D **74**, 072003 (2006), [hep-ex/0606032]
9. The MINOS Collaboration, P. Adamson et al., Phys. Rev. Lett. **101**, 131802 (2008), [hep-ex/0806.2237]
10. The T2K Collaboration, K. Abe et al., hep-ex/1106.2822
11. Particle Data Group, K. Nakamura et al., J. Phys. G **37**, 075021 (2010)
12. The MINOS Collaboration, P. Adamson et al., Phys. Rev. Lett. **106**, 181801 (2011), [hep-ex/1103.0340]
13. U. Seljak, A. Slosar, P. McDonald, JCAP **0610**, 014 (2006), [astro-ph/0604335]
14. C. Kraus et al., Eur. Phys. J. C **40**, 447 (2005), [hep-ex/0412056]

Chapter 3
Ultra-High Energy Astro-Particle Physics

Understanding the high energy Universe has been a long standing goal of astro-particle physics. Physicists have been striving to make measurements of high energy particles of astrophysical origin for the best part of a century. Particles with energies well beyond EeV have been observed for half a century [1]. However, knowledge of the identity of their sources and acceleration mechanisms is lacking. Moreover, the ability to observe such high energy particles in a reliable manner could provide tests of fundamental physics at energies far beyond those attainable by terrestrial accelerators.

Developments in astronomy have played a key role in these astro-particle physics experiments. While, at the beginning of the twentieth century, almost all astronomy was conducted via visible light, swift progress meant that by the 1960s complementary observations were being made from radio through to gamma-ray frequencies. Sensitivity at the highest energy end of this spectrum has been extended recently via the HESS [2] and VERITAS [3] observatories. These are arrays of telescopes, operating in the 100 GeV–100 TeV range, that observe Cherenkov radiation produced by air-showers that arise when gamma-rays interact in the atmosphere. Further improvements in energy range and sensitivity will be provided in the future with the construction of the Cherenkov Telescope Array [4].

Astronomy, however, is no longer confined to the detection of photons. Even while progress was being made in the broadening of the observed range of the electromagnetic (EM) spectrum, true multi-messenger astronomy had been born through the efforts of Victor Hess [5]. Hess's observation in 1912 of ionising radiation whose origin was extra-terrestrial led to the birth of cosmic-ray (CR) physics and, thus, astro-particle physics. Although astro-particle physics grew quickly as a scientific discipline, the nature of the high energy Universe is still poorly understood. Exactly how charged particles can be accelerated to $E >$ EeV is not yet known, although numerous mechanisms have been proposed.

As astro-particle physics reaches out to higher and higher energies, problems are encountered that limit the use of photons and CRs as astrophysical messengers. The increasing rarity of such high energy sources necessitates observations of increas-

M. J. Mottram, *A Search for Ultra-High Energy Neutrinos and Cosmic-Rays with ANITA-2*, Springer Theses, DOI: 10.1007/978-3-642-30032-5_3,

ingly distant objects and leads to a flux that decreases rapidly with energy. It is here that one of the biggest hurdles in astro-particle physics lies. As will be discussed later in this chapter (Sect. 3.1.5), both photon and charged messengers encounter a horizon effect at ultra-high energies. Photons pair produce e^+e^- off far infra-red (FIR) background and cosmic microwave background (CMB) radiation[1] while protons photo-pion produce with the CMB. These combined processes necessitate a new type of astrophysical messenger if the highest energy regions of the Universe, and the processes driving them, are ever to be fully understood.

The neutrino may provide physicists with a solution to this problem. As an astrophysical particle, neutrinos have only been observed in Solar emission [6] and from a single supernova explosion [7]. As purely weakly interacting particles, neutrinos could traverse cosmic distances with a negligible chance of interaction or energy loss. Furthermore, Galactic and extra-Galactic magnetic fields will not affect the paths of neutrinos. However, the weakly interacting nature of the neutrino—a virtue for high energy particles traversing the Universe—provides a challenge to observations. Vast detector volumes and novel detection methods are necessary if the observation of UHE neutrinos is ever to be realised.

3.1 Cosmic-Rays

The, as yet unobserved, UHE neutrino flux is closely linked with the production and propagation of cosmic-rays. UHE neutrinos and cosmic-rays are expected to be produced together at source. Furthermore, a significant flux of neutrinos is expected to arise from cosmic-ray interactions with the CMB as they propagate. This section covers cosmic-ray production mechanisms, potential sources and detection methods.

The cosmic-ray energy spectrum follows a power-law $dN/dE \propto E^{-\gamma}$ with relatively few features over many decades of energy, as shown in Fig. 3.1. At energies below 10^{19} eV the spectrum has been fairly well determined. In the interval 10^9 eV $\leq E \leq 10^{15}$eV the spectrum power law follows $\gamma \sim 2.7$. At the high end of this range we encounter the cosmic-ray 'knee', above which the power-law steepens to $\gamma \sim 3.1$. There are two popular explanations for this feature: PeV energies are possibly the energy upper limit of supernovae shock acceleration; Galactic magnetic fields no longer confine cosmic-rays above PeV energies. A further steepening in the cosmic-ray spectrum is observed at $\sim 4 \times 10^{17}$ eV, while a flattening feature (the 'ankle') is present at $\sim 4 \times 10^{18}$ eV. The ankle is commonly seen as the point at which the cosmic-ray flux is no longer dominated by Galactic sources, with a more diffuse, but harder, extra-Galactic flux being observed. Results from the Pierre Auger Observatory (see Sect. 3.1.3) have helped constrain the flux and composition of cosmic-rays at energies above 10^{19} eV, over which there was much uncertainty previously.

[1] $\gamma\gamma \rightarrow e^+e^-$ will begin to limit the photon horizon at 10^{14} eV.

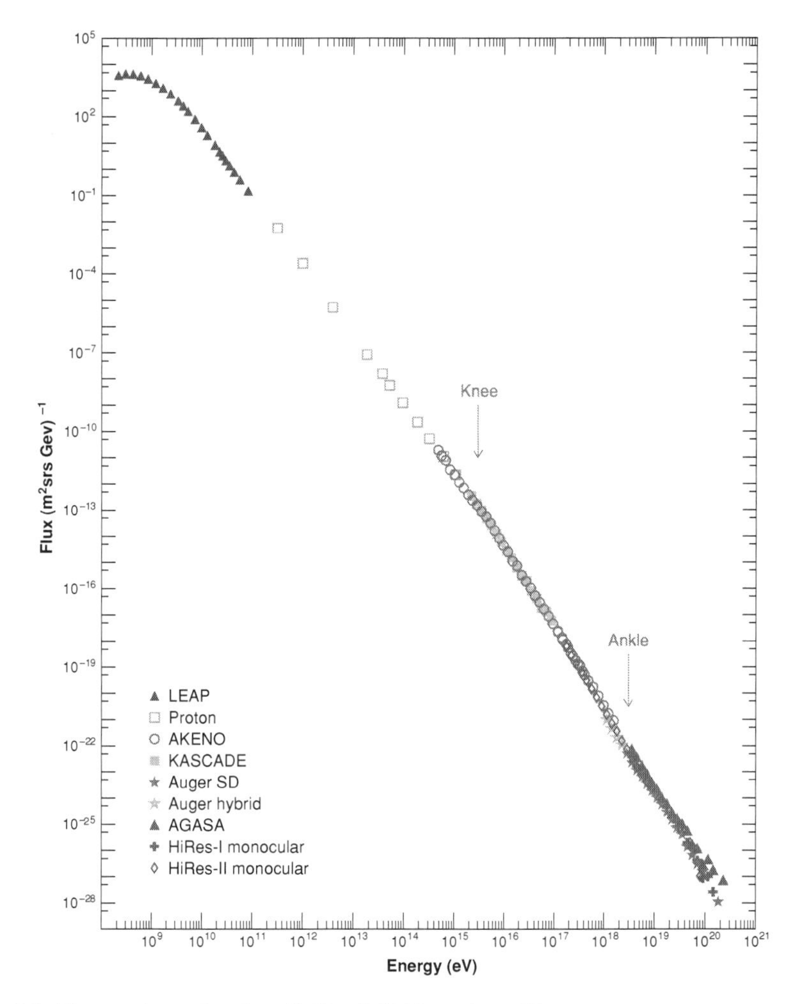

Fig. 3.1 The cosmic-ray flux from GeV to ZeV. Figure from [8]

3.1.1 Acceleration Mechanism

Acceleration through magnetic turbulence was first proposed as a mechanism for cosmic-ray acceleration by Fermi [9]. His initial proposal, known as second-order Fermi acceleration, involved clouds of dense material within the Galaxy with stronger magnetic fields than the surrounding interstellar medium. The clouds act as magnetic mirrors, a charged particle encountering an approaching cloud would be reflected and accelerated, while a charged particle encountering a receding cloud would be decelerated. Second-order Fermi acceleration was modified using the idea of shock front acceleration, with the resulting mechanism known as first-order Fermi acceleration.

In the scenario of a supersonic shock front, the shock can be viewed as a discontinuity between two regions of material: 'downstream' undisturbed gas and 'upstream' shocked, compressed, gas. In the upstream frame of reference, downstream matter will be flowing toward the upstream side, while in the downstream frame of reference, upstream matter will be flowing toward the downstream side. Therefore, matter scattered across the shock from one side is likely to scatter back across the shock. Each time the particle scatters across the shock it is accelerated in a collisionless manner by turbulent magnetic fields. Material is therefore accelerated repeatedly in a stochastic manner. For each such acceleration cycle, there is a possibility that the particle will escape on the downstream side of the shock, resulting in an expected power law energy spectrum for such an acceleration mechanism of $E \propto E^{-2}$. First-order Fermi acceleration mechanisms via supernovae shock fronts have shown to naturally give rise to a power-law energy spectrum that could replicate the observed $\gamma \sim 2.7$ [10–12].[2] First-order Fermi acceleration by supernovae remnants is now widely accepted as the means by which cosmic-rays are accelerated to energies of 10^{15} eV.

The confinement and transit of cosmic-rays is defined by their rigidity R:

$$R = \frac{pc}{Z} = r_L B \tag{3.1.1}$$

where pc is the momentum of a particle with charge Z, while r_L is the gyro-radius of the cosmic-ray in a magnetic field, B. The maximum energy attainable by a cosmic-ray in a source is dependent on the source's magnetic field, as shown by Eq. 3.1.2 [13].

$$E_{max} \sim \beta Z B R \tag{3.1.2}$$

where β is the shock velocity in terms of c and Z is the cosmic-ray's charge.

While supernovae are commonly believed to accelerate cosmic-rays to 10^{15} eV they are not believed to be able to accelerate cosmic-rays to UHE levels. Equation 3.1.2 shows that some combination of even larger source size and higher B strength are necessary if first-order Fermi acceleration is to provide observed flux of higher energy cosmic-rays. Figure 3.2 displays the size and magnetic field strengths of potential UHECR sources.

3.1.2 UHECR Acceleration

In the last two decades there has been dispute as to whether first-order Fermi acceleration is sufficient to provide the observed UHE cosmic-ray (UHECR) flux. In the 1990s, the AGASA experiment observed a surplus of cosmic-rays with $E > 10^{20}$ eV

[2] The energy spectrum produced by supernovae shock fronts is $\gamma \sim 2$. However, when combined with the confinement time in the Galaxy, a spectrum of $\gamma \sim 2.6$ is produced.

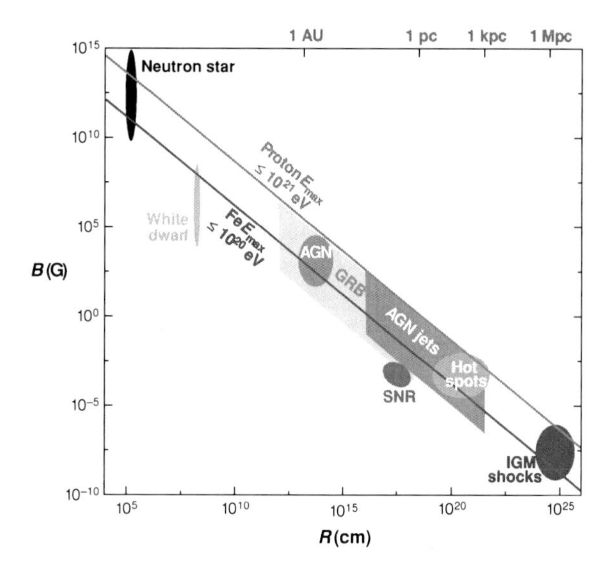

Fig. 3.2 A Hillas plot (named after [13] first proposed sources of UHECR from first-order Fermi acceleration). Sources above the red and blue lines are able to confine (and thus accelerate) iron nuclei to 10^{20} eV and protons to 10^{21} eV respectively. Figure from [14]

over a power law fit to the flux at lower energies [15]. This lead to the development of a number of so-called 'top-down' theories which proposed the decay of topological defects or super-heavy dark matter (e.g. [16, 17]) as a means of producing UHE cosmic-rays. The resulting flux would appear disjointed from the shock-front acceleration driving cosmic-ray production at lower energies. More recent data from the Auger [18] and HiRes experiments [19] have not displayed this UHECR excess. Meanwhile, limits on the UHE neutrino flux (see Sect. 3.2) also contradict top-down cosmic-ray production mechanisms. As such, these models are disfavoured.

A 'bottom-up' approach to cosmic-ray production, which utilises first-order Fermi acceleration is the current favoured model for UHECR acceleration (e.g. [20]). Of the numerous sources that have been proposed as UHECR sources, two of the most commonly suggested are shocks created by gamma-ray bursts [21] and active galactic nuclei [22], discussed below.

Gamma-Ray Bursts

Gamma-ray bursts (GRBs) are explosions of energy that, for their duration, are the most energetic events in the observable Universe, with a total energy output of up to 10^{51} erg.[3] They are observed in gamma-rays as short (from sub-second to several minutes) cosmically distant events, with afterglows often detected in X-rays. GRBs are classified as either short duration (<2 s) or long duration (>2 s). The mechanism driving the GRB is thought to differ between the two classes, with the coalescence of neutron-star or black-hole binary systems believed to result in short duration GRBs while massive star collapse leads to long duration GRBs.

[3] 1 erg = 10^{-7} J.

Active Galactic Nuclei

The term Active Galactic Nuclei (AGN) refers to galaxies with energetic phenomena at their cores that cannot be explained by stellar activity. Instead, the only way of powering the energetic processes observed is via an inferred supermassive black hole (10^7–10^9 $M_\odot$). There are a number of subclasses of AGN, including quasars, Seyfert Galaxies and BL Lacertea objects. This classification depends on factors including absolute luminosity and emission features, which are interpreted as differences in observation angles relative to the source and accretion rates of the central black hole.

Many AGN are observed to emit jets perpendicular to the plane of the galaxy and many kpc in extent. Within these jets, observations in X-ray and radio inform us that shock fronts form, it is these regions that are posited as UHECR sources. It is also possible that large magnetic fields close to AGN cores could accelerate cosmic-rays, however, these regions are thought to be optically thick and not the source of the observed UHECR flux.

Coincidence Surveys

Due to their limited horizon (discussed in Sect. 3.1.5), it is possible to conduct coincidence surveys of UHECRs with potential sources. Recent coincidence results from Auger which test the correlation of UHECR arrival directions with AGN do suggest non-isotropic arrival directions of UHECRs, however, the evidence of coincidence with AGN is tentative [23, 24]. It is possible that, while UHECR observed at Earth are produced within the expected horizon, both Galactic and inter-galactic magnetic fields are sufficient to cause larger than expected deviations in flight. Alternatively, a cosmic-ray flux at the relevant energies composed largely of heavy nuclei (as is favoured by recent Auger data [25]) would also lead to larger deflections compared to those from proton primaries.

3.1.3 Detection Methods

The flux of UHECRs is too low for direct observations to be feasible. Instead, secondary detection techniques are used that allow for larger, cost effective detectors. A cosmic-ray impinging on the Earth's atmosphere at a sufficiently high energy will cause a cascade of particles, with hadronic and electromagnetic components of the shower depending on the cosmic-ray primary. These are known as extensive air-showers (EAS), with the maximum number of particles produced proportional to the total energy of the shower.

From the 1940s to the current day, progressively larger detectors have been constructed to detect cosmic-ray-induced EAS. By the early 1960s events in the UHE regime had been observed, with the Volcano Ranch experiment detecting a cosmic-ray with $E \sim 10^{20}$ eV in 1962 [1]. Since then a number of cosmic-ray detector arrays have been constructed, with most experiments based on detection of Cherenkov radiation produced by air-shower secondaries or on observation of fluorescence light. The largest of the current generation of detectors is the Pierre-Auger observatory in Argentina [26] which combines the Cherenkov and fluorescence techniques. Auger consists of 1,600 water Cherenkov detectors, spaced 1.5 km apart over a $\sim$3,000 km^2 area. Four fluorescence observatories surround the perimeter of the water Cherenkov detector footprint, each equipped with 6 optical telescopes to cover a 180° field of view.

3.1.4 Cosmic-Ray-Induced Radio Emission

It has been known for a number of decades that EASs produce radio emission [27]. Although a number of mechanisms contribute to the overall emission, the dominant effect is that of geosynchrotron emission [28]. As the EM component of the shower develops, e^+e^- pairs will be produced, with the number of pairs proportional to the energy of the shower. These charges will experience a Lorentz force ($\vec{F}$) caused by their motion ($\vec{v}$) through the Earth's magnetic field ($\vec{B}$):

$$\vec{F} = \vec{v} \times \vec{B} \tag{3.1.3}$$

The force on e^+ will be equal and opposite to that on e^-, resulting in the separation and gyration of charges, giving rise to synchrotron radiation.

Geosynchrotron radiation will be coherent over wavelengths larger than the particle separation distances (the transverse shower size). For UHECR-induced EASs, emission will be fully coherent for frequencies below about 100 MHz. However, partial coherence of emission is still maintained in emission at higher frequencies; in fact some of the earliest observations of UHECR geosynchrotron emission were made at 500–550 MHz in the 1960s and 1970s [29, 30]. It is at the lower, coherent, frequencies that modern experimental efforts are being developed. Two European-based arrays of radio receivers lead the current generation of radio UHECR experiments; LOPES (operating in a 40–80 MHz band [31]) and CODALEMA (operating in an approximate 1–100 MHz band [32]).

3.1.5 The GZK Effect

Shortly after the Volcano Ranch UHECR observations in the 1960s, Greissen [33] and Zatsepin and Kuzmin [34] independently proposed that cosmic-rays at energies above $\sim 10^{19.5}$ eV would interact with CMB photons via a Δ^+ resonance to photo-produce pions (Eq. 3.1.4). This so-called GZK effect is expected to limit the path length of UHECRs to <100 Mpc and would lead to a noticeable steepening of the cosmic-ray energy spectrum above $10^{19.5}$ eV, known as the GZK cut-off.

If the UHECR composition were shifted to heavier nuclei, the same horizon issue would exist (albeit with a larger UHECR path length) due to photo-disintegration of the nuclei (Eq. 3.1.5). Currently the UHECR composition is a disputed matter [35, 37], although recent Auger data, shown in Fig. 3.3, indicates a mass composition that increases with energy above 10^{18} eV. Given sufficient energy, the photo-disintegrated

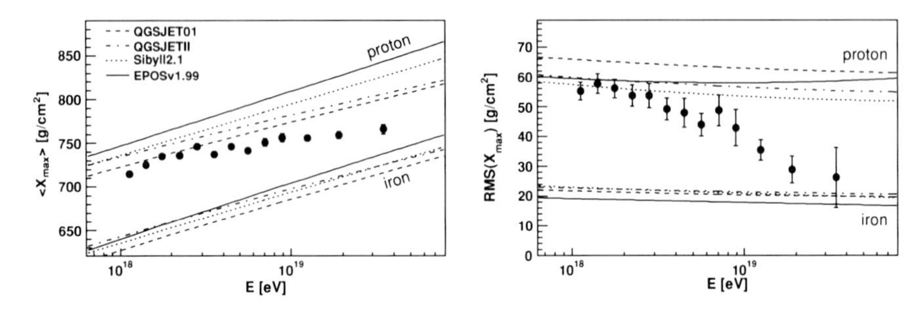

Fig. 3.3 The UHECR composition as measured by Auger. A comparison of the average depth of shower maximum, $\langle X_{max} \rangle$, and variation in depth of shower maximum $RMS(X_{max})$ is made as a function of energy from Auger data (*points*) to expected values from simulations (*lines*). Both (X_{max}) and RMS (X_{max}) measures imply a trend towards more massive cosmic-ray primaries with increasing energies when compared with various models. Figure from [35]

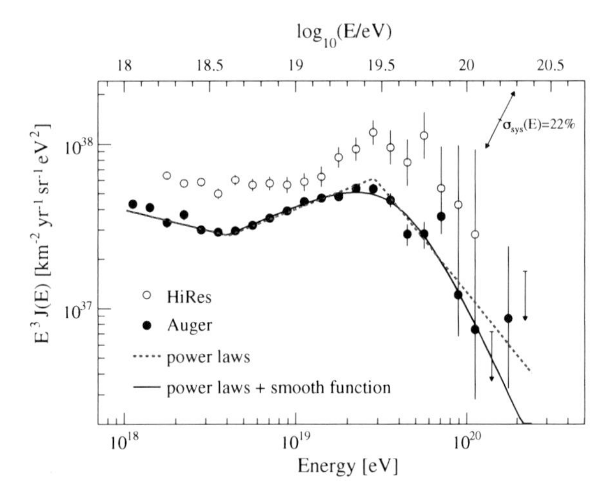

Fig. 3.4 The UHECR flux as measured by Auger. Both Auger and HiRes data [36] are shown, the "ankle" and UHECR cutoff are both clear at $E \sim 4 \times 10^{18}$ eV and $E \sim 3 \times 10^{19}$ eV respectively. Figure from [35]

nucleons could still undergo the GZK effect. However, as the average energy per nucleon is $E_{nucleus}/A$, where A is nucleon number, a higher average UHECR mass would result in the GZK cut-off shifting to higher energies. Meanwhile, a cut-off due to photo-disintegration would be apparent in the UHECR spectrum at a similar energy to the proton dominated GZK cut-off scenario.

In both proton and heavy nuclei dominated cases, the spectral steepening depends on the assumption that, at these energies, cosmic-ray sources are extragalactic. Although the most recent Auger data, along with HiRes observations, strongly support an UHECR flux cutoff at $\sim 10^{19.5}$ eV (Fig. 3.4), the previously mentioned composition studies suggest that this is due to photo-disintegration rather than an observation of the GZK effect.[4]

---

[4] It is also possible that the observed high energy cut-off is caused by the end of the energy spectrum that UHECR sources are able to generate.

$$p(E > 10^{19.5}\,\text{eV}) + \gamma_{CMB} \rightarrow \Delta^{+} \rightarrow \pi^{+}n/\pi^{0}p \tag{3.1.4}$$

$$A + \gamma_{CMB} \rightarrow (A-1) + N \tag{3.1.5}$$

3.2 UHE Neutrinos

The neutrino provides a potential solution to the horizon issues of γ-rays and cosmic-rays as astrophysical messengers. Interacting only via the weak nuclear force, neutrinos should be able to travel distances far beyond the $\sim$100 Mpc of UHECRs.

3.2.1 BZ Neutrinos

Regardless of the composition of UHECRs, it is well established that a flux beyond 10^{20} eV exists, the GZK effect is therefore expected to take place. Beresinsky and Zatsepin were the first to realise that the pions arising from the GZK process would produce UHE neutrinos via their decay chain [38], resulting in a 'guaranteed' flux of cosmogenic, or 'BZ', neutrinos (Eq. 3.2.1). The flux of BZ neutrinos is expected to be significant, dominating over other sources in the UHE regime if the UHECR composition is predominantly protons. Figure 3.5 shows one such prediction of the UHE neutrino flux.

In the case of a UHECR flux comprising more massive nuclei, photo-disintegration will also result in UHE neutrinos being generated via neutron decay (Eq. 3.2.2). This heavy mass UHECR composition, which is supported by data from Auger, would significantly reduce the BZ neutrino flux in the UHE regime, as shown in Fig. 3.6.

$$\begin{aligned} \pi^{+} &\rightarrow \mu^{+} + \nu_{\mu} \\ &\quad \hookrightarrow e^{+} + \nu_{e} + \bar{\nu}_{\mu} \end{aligned} \tag{3.2.1}$$

$$n \rightarrow p^{+} + e^{-} + \bar{\nu}_{e} \tag{3.2.2}$$

Note that the neutrino flux produced via the GZK effect seen in equation 3.2.1 would consist of a ratio of flavour states $\nu_e : \nu_\mu : \nu_\tau$ of $\frac{1}{3} : \frac{2}{3} : 0$, with both ν_μ and $\bar{\nu}_\mu$ produced. Meanwhile, the flux produced via photo-disintegration and neutron decay would give rise only to $\bar{\nu}_e$.

Even with a pure proton UHECR composition, high energy neutrons will be produced via the GZK process (Eq. 3.1.4). The BZ neutrino flux is therefore expected to combine neutrinos from both Eqs. 3.2.1 and 3.2.2.

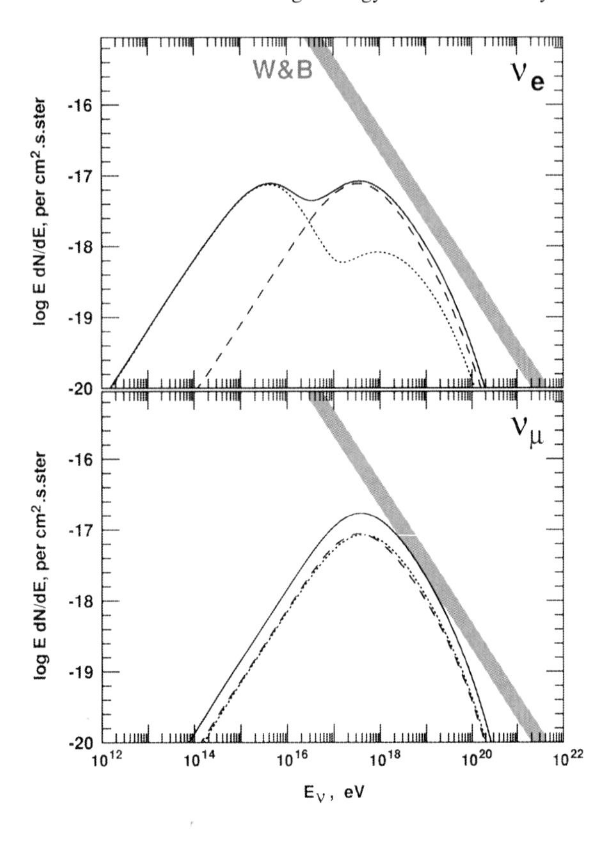

Fig. 3.5 Fluxes of ν_e (*top*) and ν_μ (*bottom*) neutrinos predicted by [41]. Neutrino fluxes are denoted by *dashed lines*, anti-neutrinos by *dotted lines*, sum total by *solid lines*. The Waxman-Bahcall limit for neutrino production within cosmic-ray sources is shown by the shaded band in each case. The double peak displayed by the ν_e spectra is caused by neutrinos from pion decay (higher energies) and neutron decay (lower energies)

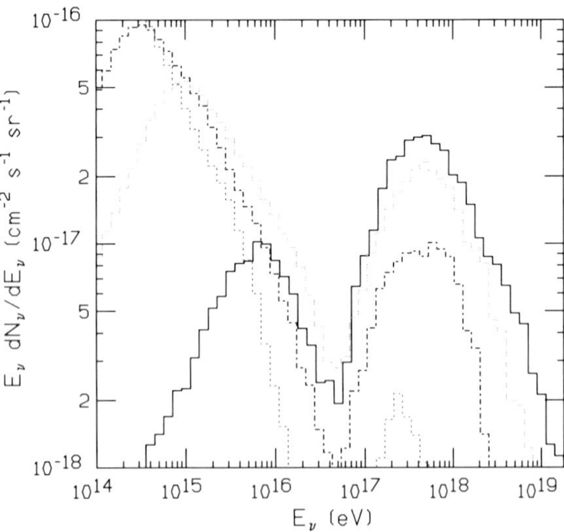

Fig. 3.6 Expected fluxes of BZ neutrinos for protons (*black*, *solid*), ^{4}He (*green*, *dashed*), ^{16}O (*red*, *dotted*) and ^{56}Fe (*blue*, *dots*). Figure from [42]

3.2.2 The Waxman-Bahcall Limit

Using the first-order Fermi shock-front production mechanism of UHECR production within extra-Galactic sources, Waxman and Bahcall postulated a model independent flux limit on UHE neutrinos produced within these sources [39, 40]. Waxman and Bahcall argued that to produce the observed UHECR flux, the mean free path of protons in the source must be low enough to allow the cosmic-rays' escape. However, some fraction of the UHECR will photo-pion produce within the source, imparting a fraction of their energy onto the neutrinos from the resulting pion decay. Using the observed UHECR flux, it is therefore possible to calculate a limit on the direct neutrino flux from UHECR sources.

The value of the Waxmax-Bahcall limit is $E_\nu^2 \Phi_\nu < 2 \times 10^{-8}\,\mathrm{GeV/cm^2\,s\,sr}$. This relies on the first-order Fermi acceleration mechanisms being applicable to the energy range being considered. Waxman and Bahcall note that, in the case of an UHE source that is optically thick to protons such as the regions close to an AGN core, it would be possible for the neutrino flux to exceed the Waxman-Bahcall limit. However, they argue that such sources cannot explain the observed UHECR flux.

3.2.3 Particle Physics with UHE Neutrinos

Measurements of the UHE neutrino flux (or placing limits on it) will provide evidence as to whether the GZK effect takes place, as well as providing insight on UHECR acceleration mechanisms. UHE neutrino experiments are further motivated from a particle physics perspective. For instance, the centre of mass energy of a 10^{19} eV neutrino interaction with a non-relativistic proton is $\sim$137 TeV; an order of magnitude higher than the collision energy at the LHC.

If a flux of UHE neutrinos were established, measurements of flavour composition could allow for tests of particle physics models of neutrino mixing. While a one- or two-flavour flux is expected to arise from the two possible production mechanisms (Eqs. 3.2.2 and 3.2.1 respectively), on observation at Earth the flux is expected to be maximally mixed ($\nu_e : \nu_\mu : \nu_\tau$ of close to $1 : 1 : 1$). Both the oscillation baseline ($\sim$100s Mpc) and neutrino energies (EeV–ZeV) being considered are far in excess of current neutrino oscillation experiments. Such observations would allow for tests of the current oscillation model, as well as providing a probe of beyond Standard Model physics, for example Lorentz and CPT violation, which would lead to modifications of oscillations only detectable at the highest energies and cosmic distances [43].

Finally, measuring the UHE neutrino cross section provides a test for extra-dimensions. At 10^{19} eV, the neutrino cross section is expected to be approximately $0.3 \times 10^{-31}\mathrm{cm}^2$, leading to an interaction length of $O(100)$ km in rock [44–46]. UHE neutrinos will therefore be absorbed by the Earth, allowing for a measurement of the cross section via analysis of UHE neutrino arrival directions in detectors.

Of course, tests of oscillation parameters, neutrino cross-sections and exotic physics with UHE neutrinos would require a statistically significant sample of events, which is currently a far off goal.

3.2.4 Detection of UHE Neutrinos

Despite suggestions that there is a 'guaranteed' flux of the particles, no detection of UHE neutrinos have been made to date. A number of experiments designed for this purpose have been constructed, utilising a range of detection concepts.

Cosmic-ray air-shower detectors are sensitive not only to cosmic-ray-induced EASs, but also those initiated by UHE neutrinos. By assuming any UHE particle other than a neutrino would interact within the Earth's atmosphere well before reaching the Earth's surface, experiments such as Auger can differentiate neutrinos from other primaries by searching for highly inclined, or very deep, showers. The progenitor particle will have traversed many more km of atmosphere than would be possible for any primary other than a neutrino. Auger has produced a limit on UHE ν_τ [47] by searching for Earth skimming air-showers (i.e. showers with an inferred direction of development that is either slightly inclined from *below* the detector or that points towards nearby mountain ranges). These showers are a signature of ν_τ that have interacted in the Earth via charged–current processes, producing a collinear τ particle that decays in flight, resulting in an EAS.

Large scale (km^3) optical detectors aim to observe the Cherenkov emission of a secondary lepton from neutrino interactions. The only such detector currently operating is IceCube, situated at the South Pole [48]. A further experiment, km3Net, is planned for construction in the Mediterranean [49]. Both IceCube and km3Net are based on or around smaller concept experiments, with the IceCube array surrounding its predecessor, AMANDA [50], and km3Net intended as the successor to the ANTARES experiment [51]. These experiments primarily search for lower energy cosmic neutrinos in the TeV–PeV range. In order to remove cosmic-ray-induced μ backgrounds, IceCube and AMANDA search for neutrinos passing through the Earth ('up-going' neutrinos). Optical Cherenkov experiments are also sensitive to $>$PeV neutrinos when searching for above horizon ('down-going') neutrinos. However, size constraints severely limit exposure at $E_\nu > 10^{18}$ eV, as the optical attenuation length in ice and water mean that any significant increase in the volume of optical detectors would require many more detector modules and, as such, would be prohibitively expensive.

UHE neutrino interactions in sufficiently dense media are expected to cause rapid thermal expansion that could be detected acoustically. Given a sufficiently large attenuation length, acoustic experiments could be constructed in arrays of many km^3. However, currently only proof-of-concept experiments have been constructed. Acoustic detection of neutrinos has been investigated using ice and water as interaction media, through the ACoRNE [52], AMADEUS [53], Lake Baikal [54], SAUND [55] and SPATS [56] experiments.

The most promising detection technique for UHE neutrino detection is that of coherent radio emission from neutrino-induced shower secondaries, described in the following section.

3.2.5 Radio Detection of UHE Neutrinos

Gurgen Askaryan proposed in the 1960s that an ultra high energy neutrino interaction in a dielectric medium would give rise to coherent Cherenkov radiation in the radio regime [57, 58]. Consider an UHE neutrino traversing some suitable interaction medium. This neutrino is able to interact via either charged—or neutral—current processes, both of which will cause hadronic showers through nuclear recoil which, in turn, will give rise to an EM particle shower. In the case of a $\nu_i e^-$ interaction, the interacting electron will also initiate an EM shower. Over the development of this EM cascade, a charge imbalance will develop. Electrons will be up-scattered into the shower via the inverse Compton effect, while positrons annihilate out of the shower. Askaryan predicted that the total charge imbalance of the shower would be around 20 %.

This EM shower will have a transverse dimension that is characterised by the Moliére radius, R_M, that is dependent on the interaction medium:

$$R_M = X_0 21.\,\mathrm{MeV}/E_c \tag{3.2.3}$$

where X_0 is the radiation length[5] and E_c is the critical energy[6] of the interaction medium. For ice, $X_0 \sim 40\,\mathrm{cm}$ and $E_c \sim 56\,\mathrm{MeV}$, leading to $R_M = O(10)$.

The shower will develop over several metres within the ice, though the instantaneous longitudinal dimension of the particle bunch (that is, the dimension along the shower axis) will be $O(1)\,\mathrm{cm}$. The total energy, W_{tot}, produced by a single-charged particle emitting Cherenkov radiation in a frequency range $\nu_{min} - \nu_{max}$ over a track length L is:

$$W_{tot} = \frac{\pi \alpha h}{c} \left(1 - \frac{1}{n^2\beta^2}\right) L \left(\nu_{max}^2 - \nu_{min}^2\right) \tag{3.2.4}$$

where $\alpha \sim 1/137$ is the fine structure constant, h is Planck's constant, c is the speed of light, n is the index of refraction of the medium and β is the particle's velocity in terms of c. At wavelengths larger than the transverse shower size, Cherenkov

[5] The radiation length is the length over which a particle will lose all but $1/e$ of its energy.

[6] The critical energy is the energy at which an electron's rate of energy loss due to ionisation is equal to the rate of energy loss due to bremsstrahlung radiation. Ionisation interactions do not create any new particles, while bremsstrahlung energy losses result in a photon, which may proceed to pair produce, thus adding to the total number of particles in the shower. Therefore, at $E < E_c$, particles will not be added to the shower.

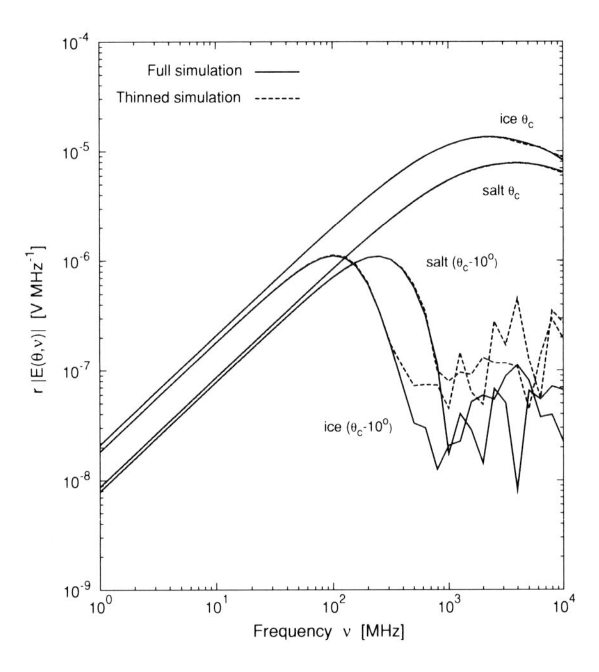

Fig. 3.7 Askaryan emission from two simulations for a 100 TeV primary neutrino interacting in either ice or salt. As the angle of observation moves away from the Cherenkov angle, both the amplitude of the electric field and the high frequency cut-off decrease. Figure from [59]

radiation emitted by particles in the shower will become coherent in nature. Once in the coherent regime, the power of the Cherenkov emission from a shower scales as $P \propto N^2$, or equivalently $W_{tot} \propto N^2$, where N is the number of super-luminal charges. Figure 3.7 shows the expected frequency spectrum for Askaryan radiation in ice and salt.

The Askaryan effect was experimentally confirmed at SLAC in 2001 by sending picosecond pulses of GeV photons into a sand target, resulting in EM showers that developed over a number of metres [60]. This initial measurement was followed by a comparable measurement using salt [61]. A further measurement was conducted using an ice target and picosecond bunches of GeV electrons [62]. Figure 3.8 shows the dependence of emission strength on frequency and shower energy from the ice-target experiment. This last observation of the Askaryan effect was conducted using the ANITA payload as the experimental radio detector.

Figure 3.9 demonstrates that the Askaryan emission measurements made by [60–62] are consistent with a bimodal signal that is sub-ns in duration. However, the expected signal arising from an UHE neutrino-induced cascade may differ significantly from those caused by bunches of lower energy particles due to an expected longer longitudinal shower profile. With bremsstrahlung and pair production interaction distances on a comparable scale to the inter-atomic spacing of the media in which the shower develops, the Landau-Pomeranchuk-Migdal (LPM) effect [63–65] can result in an increase of mean free path of photons and electrons in an UHE-induced shower. This, in turn, increases the longitudinal length of a shower and resulting

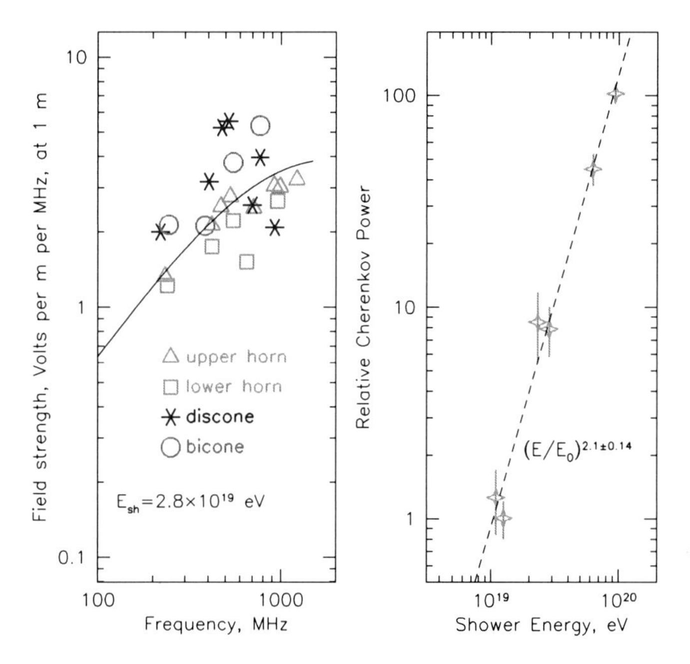

Fig. 3.8 *Left* Askaryan pulse field strength as measured in ice as a function of frequency. *Right* Demonstration of the quadratic dependence of emission strength on shower energy. Figure from [62]

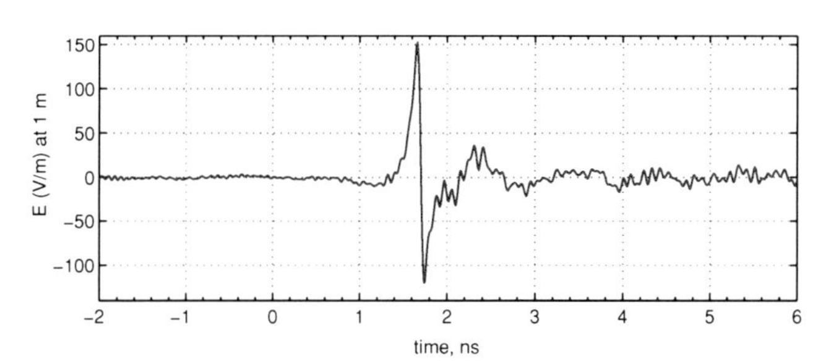

Fig. 3.9 Askaryan pulse field strength as measured in salt. Figure from [61]

Askaryan signals appear less impulsive in nature, with higher energy primaries leading to longer duration signals [66].

A number of experiments were beginning to utilise radio techniques in UHE neutrino searches around the same time as the initial experimental confirmations of the Askaryan effect. The FORTE satellite was launched in 1997 with the intention of testing nuclear detonation detection. Data from FORTE has also been used to set a limit on the UHE neutrino flux using radio observations of the Greenland ice-sheet over a 30–300 MHz bandwidth [67].

The Goldstone Lunar UHE Neutrino Experiment (GLUE) made use of the JPL/NASA Deep Space Network antennas at Goldstone, California, to observe the lunar regolith for signs of Askaryan radiation [68]. The experiment searched for radio pulses from the moon that were $\leq$10 ns in duration, requiring coincidence in observations between two antennas, separated by 22 km, to reject backgrounds.

The Radio Ice Cherenkov Experiment (RICE) is an embedded experiment at the South Pole [69] that searches for Askaryan emission from neutrino interactions in the ice surrounding the detector. RICE consists of 18 radio receivers operating in a frequency range of 100 MHz–1 GHz. The antennas are deployed in a 800 m^3 cube-shaped volume, at a depth of 100–300 m. RICE is situated 600 m above the AMANDA optical neutrino telescope.

The exposure of each of the mentioned experiments is limited in some manner, through exposure time (GLUE), observable ice volume (RICE) or detector sensitivity (FORTE). As a result, none of the experiments has been able to place limits on the neutrino flux that are able to constrain non-exotic flux models. The ANITA experiment is ideally placed to improve significantly on the limits of each of the three experiments, covering the energy range of all three by combining dedicated sensitive observations with a vast interaction volume. In the future, the Askaryan Radio Array (ARA) will provide a radio extension to the IceCube experiment [70], with the potential to improve on the sensitivity of ANITA.

References

1. J. Linsley, Phys. Rev. Lett. **10**, 146 (1963)
2. The HESS Collaboration, J.A. Hinton, New Astron. Rev. **48**, 331 (2004), [astro-ph/0403052]
3. J. Holder et al., Status of the VERITAS observatory, in *American Institute of Physics Conference Series*, vol. 1085, (2008) pp. 657–660, [astro-ph/0810.0474]
4. The CTA Consortium, astro-ph/1008.3703
5. V. Hess, *Nobel Lectures, Physics 1922–1941* (Elsevier, Amsterdam, 1965)
6. B.T. Cleveland et al., Astrophys. J. **496**, 505 (1998)
7. K. Hirata et al., Phys. Rev. Lett. **58**, 1490 (1987)
8. J.J. Beatty, S. Westerhoff, Ann. Rev. Nucl. Part. Sci. **59**, 319 (2009)
9. E. Fermi, Phys. Rev. **75**, 1169 (1949)
10. W.I. Axford, E. Leer, G. Skadron, The acceleration of cosmic rays by shock waves, in *International Cosmic Ray Conference* vol. 11 (1977), 132–137
11. A.R. Bell, Mon. Notices R. Astron. Soc. **182**, 147 (1978)
12. R.D. Blandford, J.P. Ostriker, Astrophys. J. **221**, L29 (1978)
13. A.M. Hillas, Ann. Rev. Astron. Astrophys. **22**, 425 (1984)
14. K. Kotera, A.V. Olinto, Ann. Rev. Astron. Astrophys. **49**, 119 (2011)
15. M. Takeda et al., Astropart. Phys. **19**, 447 (2003), [astro-ph/0209422]
16. V. Berezinsky, M. Kachelrieß, A. Vilenkin, Phys. Rev. Lett. **79**, 4302 (1997)
17. V. Berezinsky, A. Vilenkin, Phys. Rev. Lett. **79**, 5202 (1997)
18. The Pierre Auger Collaboration, J. Abraham et al., Phys. Lett. B685, 239 (2010), [astro-ph/1002.1975]
19. The High Resolution FlyGs Eye Collaboration, R.U. Abbasi et al., Phys. Rev. Lett. **100**, 101101 (2008)
20. A.D. Erlykin, A.W. Wolfendale, J. Phys. **G31**, 1475 (2005), [astro-ph/0510016]

21. E. Waxman, Phys. Rev. Lett. **75**, 386 (1995), [astro-ph/9505082]
22. R.J. Protheroe, A.P. Szabo, Phys. Rev. Lett. **69**, 2885 (1992)
23. The Pierre Auger Collaboration, P. Abreu et al., Astropart. Phys. **34**, 314 (2010), [astro-ph/1009.1855]
24. L.J. Watson, D.J. Mortlock, A.H. Jaffe, astro-ph/1010.0911
25. The Pierre Auger Collaboration, J. Abraham et al., astro-ph/0906.2319
26. The Pierre Auger Collaboration, J. Abraham et al., Nucl. Instrum. Meth. **A523**, 50 (2004)
27. J.V. Jelley et al., Nature **205**, 327 (1965)
28. F.D. Kahn, I. Lerche, R. Soc. Lond. Proc. Ser. A **289**, 206 (1966)
29. D.J. Fegan, D.M. Jennings, Nature **223**, 722 (1969)
30. D.J. Fegan, P.P. O'Neill, Nature **241**, 126 (1973)
31. The LOPES Collaboration, H. Falcke et al., Nature **435**, 313 (2005), [astro-ph/0505383]
32. The CODELAMA Collaboration, D. Ardouin et al., Astropart. Phys. **31**, 192 (2009), [astro-ph/0901.4502]
33. K. Greisen, Phys. Rev. Lett. **16**, 748 (1966)
34. G.T. Zatsepin, V.A. Kuzmin, JETP Lett. **4**, 78 (1966)
35. The Pierre Auger Collaboration, J. Abraham et al., Phys. Rev. Lett. **104**, 091101 (2010), [astro-ph/1002.0699]
36. The High Resolution FlyGs Eye Collaboration, R.U. Abbasi et al., Astropart. Phys. **32**, 53 (2009), [astro-ph/0904.4500]
37. The High-Resolution Fly's Eye Collaboration, R.U. Abbasi et al., Phys. Rev. Lett. **104**, 161101 (2010), [astro-ph/0910.4184]
38. V.S. Beresinsky, G.T. Zatsepin, Phys. Lett. B **28**, 423 (1969)
39. E. Waxman, J.N. Bahcall, Phys. Rev. D **59**, 023002 (1999), [hep-ph/9807282]
40. J.N. Bahcall, E. Waxman, Phys. Rev. D **64**, 023002 (2001), [hep-ph/9902383]
41. R. Engel, D. Seckel, T. Stanev, Phys. Rev. D **64**, 093010 (2001), [astro-ph/0101216]
42. D. Hooper, A. Taylor, S. Sarkar, Astropart. Phys. **23**, 11 (2005), [astro-ph/0407618]
43. D. Hooper, D. Morgan, E. Winstanley, Phys. Rev. D **72**, 065009 (2005), [hep-ph/0506091]
44. R. Gandhi, C. Quigg, M.H. Reno, I. Sarcevic, Astropart. Phys. **5**, 81 (1996), [hep-ph/9512364]
45. R. Gandhi, C. Quigg, M.H. Reno, I. Sarcevic, Phys. Rev. D **58**, 093009 (1998), [hep-ph/9807264]
46. A. Connolly, R.S. Thorne, D. Waters, hep-ph/1102.0691
47. The Pierre Auger Collaboration, J. Abraham et al., Phys. Rev. D **79**, 102001 (2009)
48. F. Halzen, S.R. Klein, Rev. Sci. Instrum. **81**, 081101 (2010), [astro-ph/1007.1247]
49. The KM3NeT Collaboration, M. de Jong, Nucl. Instrum. Meth. **A623**, 445 (2010)
50. The AMANDA Collaboration, E. Andres et al., Astropart. Phys. **13**, 1 (2000), [astro-ph/9906203]
51. The ANTARES Collaboration, J. Aguilar et al., Phys. Lett. B **696**, 16 (2011), [1011.3772]
52. The ACoRNE Collaboration, S. Danaher, J. Phys. Conf. Ser. **81**, 012011 (2007)
53. The AMADEUS Collaboration, R. Lahmann, Nucl. Instrum. Methods **A604**, S158 (2009), [astro-ph/0901.0321], * Brief entry *
54. V. Aynutdinov et al., astro-ph/0910.0678
55. N. Kurahashi, J. Vandenbroucke, G. Gratta, Phys. Rev. D **82**, 073006 (2010), [hep-ex/1007.5517]
56. S. Boeser et al., astro-ph/0807.4676
57. G.A. Askaryan, JETP **14**, 441 (1962)
58. G.A. Askaryan, JETP **21**, 658 (1965)
59. J. Alvarez-Muniz, C. James, R. Protheroe, E. Zas, Astropart. Phys. **32**, 100 (2009)
60. D. Saltzberg et al., Phys. Rev. Lett. **86**, 2802 (2001), [hep-ex/0011001]
61. P.W. Gorham et al., Phys. Rev. D **72**, 023002 (2005), [astro-ph/0412128]
62. The ANITA Collaboration, P.W. Gorham et al., Phys. Rev. Lett. **99**, 171101 (2007), [hep-ex/0611008]
63. L.D. Landau, I. Pomeranchuk, Dokl. Akad. Nauk Ser. Fiz. **92**, 535 (1953)
64. L.D. Landau, I. Pomeranchuk, Dokl. Akad. Nauk Ser. Fiz. **92**, 735 (1953)

65. A.B. Migdal, Phys. Rev. **103**, 1811 (1956)
66. J. Alvarez-Muniz, A. Romero-Wolf, E. Zas, Phys. Rev. D **81**, 123009 (2010), [astro-ph/1002.3873]
67. N.G. Lehtinen, P.W. Gorham, A.R. Jacobson, R.A. Roussel-Dupré, Phys. Rev. D **69**, 013008 (2004)
68. P.W. Gorham et al., Phys. Rev. Lett. **93**, 041101 (2004)
69. The RICE Collaboration, I. Kravchenko et al., Astropart. Phys. **19**, 15 (2003), [astro-ph/0112372]
70. P. Allison et al., astro-ph/1105.2854

Chapter 4
The Antarctic Impulsive Transient Antenna

The Antarctic Impulsive Transient Antenna (ANITA) was conceived to be the most sensitive instrument for the detection of UHE neutrinos ($E_\nu > 10^{18}$ eV) to date. A balloon-borne array of radio antennas, ANITA observes the Antarctic ice sheet and searches for the impulsive radio emission expected to arise when an UHE neutrino interacts in the ice.

Figure 4.1 gives a schematic view of the ANITA concept. A neutrino interacting in the ice will give rise to coherent Cherenkov radio emission. For certain shower directions, some of this emission may pass through the ice-air boundary, and be observed by ANITA. The signature of this emission will be a short duration (impulsive) signal.

Antarctica provides an ideal environment for the ANITA experiment, with a vast volume of radio transparent interaction material and relatively few sources of radio frequency (RF) backgrounds. Long duration balloon (LDB) flights have typical altitudes of 36 km. The resulting horizon distances of up to 700 km provide ANITA with $\sim 1.5 \times 10^6$ km^2 of viewable Antarctic ice at any one time. With attenuation lengths at radio frequencies in Antarctic ice measured at $O(1{,}000)$ m [1], this means that ANITA observes an interaction volume of $> 10^6$ km^3.

The ANITA experiment has completed two science flights, with some modifications made between the first and second flight. Where there are differences between the two flights, the terms ANITA-1 and ANITA-2 will be used to remove ambiguity, where ANITA appears, it refers to features present in both flights. A detailed description of the ANITA-1 instrument appears in [2].

4.1 Results from ANITA-1

The analysis of ANITA-1 data returned 16 candidate EAS geosynchrotron events [3]. Figure 4.2 shows the results of an analysis that compared the observed radio emission polarisation with the inclination of the magnetic field at the projected location of the shower maximum. The correlation observed between these two parameters

M. J. Mottram, *A Search for Ultra-High Energy Neutrinos and Cosmic-Rays with ANITA-2*, Springer Theses, DOI: 10.1007/978-3-642-30032-5_4,

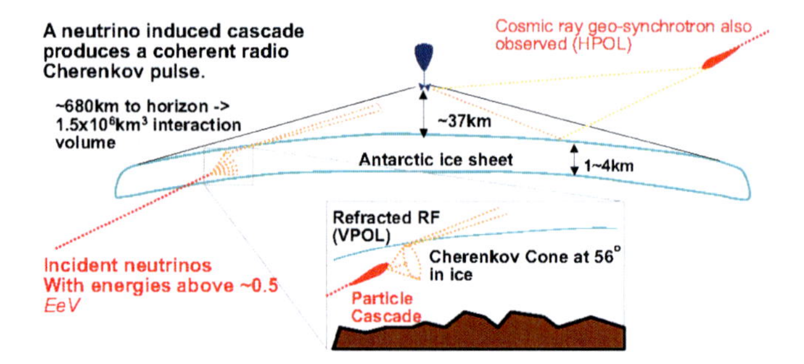

Fig. 4.1 The ANITA detection concept; Earth-skimming neutrinos interact in the Antarctica ice sheet producing downward-pointing (in the instrument's reference view) vertically polarised event signatures, while extensive air showers result in horizontally polarised signals, with observations of either direct or reflected emission possible

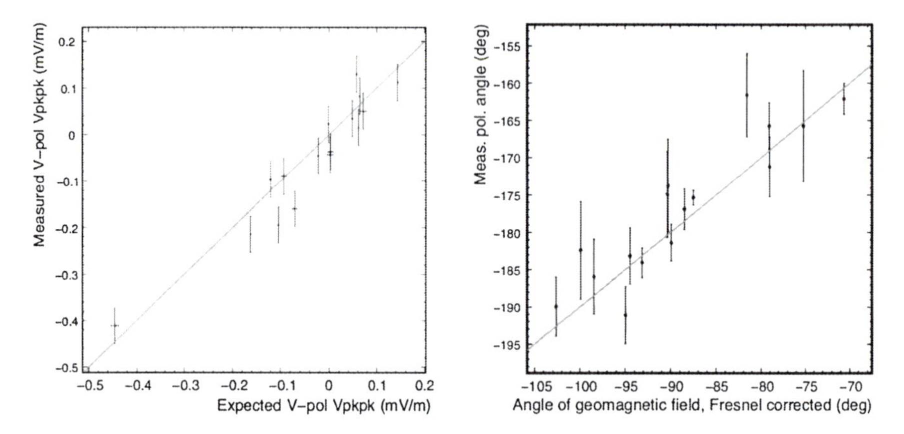

Fig. 4.2 UHECR identified events from ANITA-1. *Left*: measured vs. expected $V_{peak-peak}$ in the vertical polarisation, *right*: measured event polarisation vs. $\vec{B}$-Field inclination at the projected UHECR interaction location. Note that the green lines are expected values, not fits. Figure from [4]

confirmed that the radio emission observed did arise from UHECR interactions. Current understanding of geosynchrotron emission and air-shower dynamics results in larger uncertainties in the calculation of primary UHECR energy from ANITA data compared with data from fluorescence and optical Cherenkov experiments such as Auger and HiRes. However, best estimates of the energy of the 16 isolated events observed by ANITA place the energy of each of the primary cosmic-rays at $E_{CR} > 10^{18}$ eV. This represents a significant advance in the sample of UHECR events observed through radio techniques and demonstrates a promising avenue of research for a possible future ANITA flight.

ANITA-1 observed no statistical evidence of emission arising from UHE neutrino interactions. Results from the first flight placed a limit on cosmic neutrino flux [3, 5],

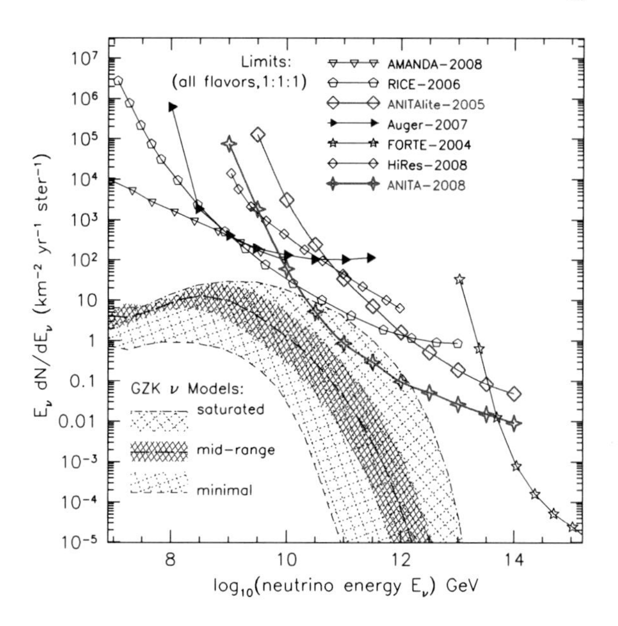

Fig. 4.3 Limit on the UHEν flux from the first flight of ANITA. Other limits are plotted for AMANDA [8], ANITA-lite [9], Auger [10], FORTE [11], HiRes [12] & RICE [13]. Figure from [3]

shown in Fig. 4.3. This limit was the most stringent in its energy range, until analysis of data from the second flight further constrained the flux [6, 7].

4.2 Experiment Overview

Figure 4.4 shows the ANITA-2 payload prior to flight. ANITA-2 detects RF signals with an omni-directional array of 40 dual-polarised antennas. The instrument's systems are optimised to operate over a 200–1200 MHz band. The instrument is designed to detect electric fields of $O(1\,\text{mV/m})$, meaning the introduction of thermal noise must be minimised.

4.2.1 Expected Signals

UHE neutrinos

As described in Sect. 3.2.5, UHE neutrino interactions are expected to give rise to radio emission that is brief in duration ($O(1\,\text{ns})$) and broadband in nature. Although UHE neutrinos are expected to traverse distances $\gg 10^3$ Mpc of the intergalactic medium (IGM) uninterrupted, the Earth is expected to be opaque to them. Extrapolated Standard Model neutrino cross-sections scale with energy, such that

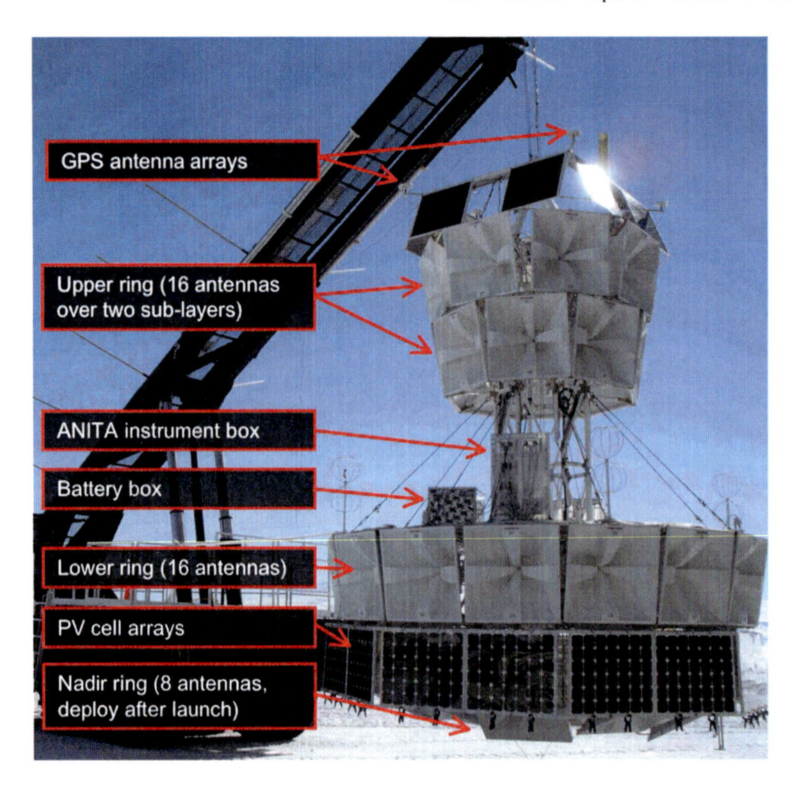

Fig. 4.4 The ANITA-2 payload suspended from the launch vehicle prior to flight

at $E_\nu \geq$ EeV neutrino interaction lengths are expected to be $O(100)$ km water-equivalent [14, 15]. The result of this interaction length is that any 'up-going' neutrino (impinging on the opposite side of the Earth, but traveling towards a detector) will be absorbed well before it is able to be detected. Meanwhile, the majority of 'down-going' interacting neutrinos will emit radiation away from the instrument.[1] ANITA is therefore optimally sensitive to 'Earth-skimming' neutrinos that are incident at highly inclined angles with respect to the Earth's local zenith angle.

The Cherenkov emission angle is given by $\cos(\theta_C) = \frac{1}{n\beta}$, where n is the refractive index of the medium and β is the particle velocity in terms of the speed of light. For Antarctic ice ($n_{ice} \sim 1.758$ in the RF regime) this gives $\theta_C \sim 55.8°$. For Earth-skimming showers, only the top portion of the resultant Cherenkov emission cone will be incident on the ice-air boundary at angles below the critical angle. Figure 4.5 shows that neutrino-induced Askaryan emission observable by ANITA will therefore be predominantly vertically polarised.

---

[1] There is the possibility that interactions of some down-going neutrinos may be observable by ANITA through reflected signals, particularly if the interactions take place within one of the two large ice-shelfs in Antarctica where the coefficient of reflection from the ice-sea-water boundary is very high and the ice thickness is <1 km [16].

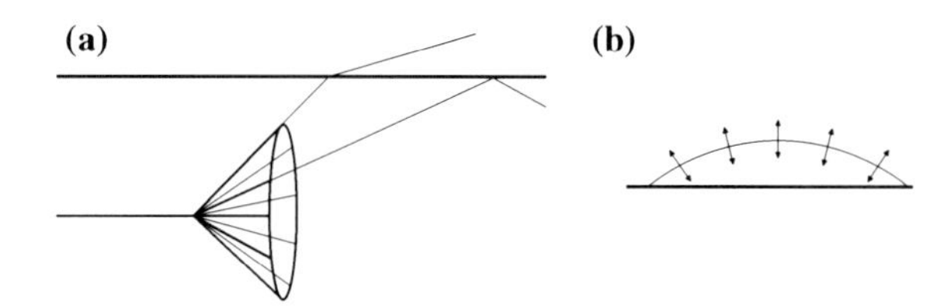

Fig. 4.5 **a**: Schematic of Askaryan radiation from a neutrino interacting in the ice, with the top portion of the Cherenkov cone escaping the ice. **b**: The top portion of the Cherenkov cone that escapes the ice will be predominantly vertically polarised

Cosmic-Rays

As discussed in Sect. 3.1.4, ANITA-1 observed a number of UHE cosmic-rays (UHECRs) through geosynchrotron emission. When an UHECR is incident on the Earth's atmosphere it will interact, the resulting extensive air-showers (EASs) will develop over kilometre scales. A Lorentz force, caused by the Earth's magnetic field, results in the separation and gyration of electrons and positrons which then produce radio emission (geosynchrotron). This emission at very southerly latitudes will be predominantly horizontally polarised, due to the local magnetic field orientation.

The ANITA-1 observations showed that ANITA is able to detect both direct and reflected UHECR geosynchrotron emission, as displayed in Fig. 4.1. UHECRs will interact below the altitude at which ANITA flies. For emission to be observed directly, the EAS must develop in a manner that the emission never reaches the ground, leading to an angular acceptance of $\sim 1°$. Meanwhile, reflected events have no such constraint. The aperture for reflected events is therefore significantly greater than that for direct events.

4.2.2 Design Obstacles

A number of compromises, discussed throughout this chapter, were necessary for ANITA to meet dimension, weight, budget and power constraints. To utilise the maximum volume of observable ice, ANITA required full azimuthal coverage, along with maximal coverage over the zenith angles in which the Antarctic ice-sheet is in view. However, size constraints were placed on the total detector size by the balloon launching process; the antenna array had to fit within a diameter of <10 m, while the total height of the instrument was limited to a similar size.

ANITA must be able to record radio waveform data for every event it records to allow analysis of events to reconstruct the source locations of radio emission. This waveform data must be digitised for storage during flight. However, power and memory limitations meant that commercially available digitising chips were not viable at the time of ANITA's design, therefore the ANITA data acquisition hardware was based on a specifically designed application-specific integrated circuit (ASIC) known as the Large Analogue Bandwidth Recorder And Digitizer with Ordered Readout (LABRADOR) chip [17].

4.2.3 Gondola Design and Power Systems

The structure of the ANITA-2 payload, whose dimensions were at the limit imposed by the Columbia Scientific Balloon Facility (CSBF) Long Duration Balloon (LDB) constraints, was designed to be both lightweight and easily constructed and dismantled to facilitate ease of transportation. The gondola itself (Fig. 4.4) was made of hollow aircraft-grade aluminium alloy tubing connected via a range of joints, with no single part larger than a couple of metres in length.

Power was provided by an omnidirectional array of photovoltaic (PV) cells. The PV cell outputs led into a charge controller, which distributed power to the payload at a steady 24-V, with DC-DC converters implemented to provide the range of voltages (+5 V, +12 V, −12 V, +3.3 V, +1.5 V) required by the various ANITA-2 systems. A farm of lead-acid batteries, housed on the payload deck, were also connected to the charge controller. The charge controller allowed ANITA-2's subsystems to draw additional power from the battery farm when required, or, in times of high power output from the PV cells, allowed the battery farm to recharge with the excess power being produced.

4.2.4 CSBF Support

All CSBF LDB flights carry a science instrumentation package (SIP) that is powered and operated independently from the experiment hardware being flown. The SIP is used for all telemetry and flight control (e.g. ballast control and flight termination), as well as providing separate navigational information from that used by the scientific instrument.

The SIP was situated on the main deck of ANITA-2, on the opposite side of the payload to the ANITA instrument box. As it is vital that local electromagnetic interference (EMI) is kept at the minimum achievable level, the CSBF SIP was placed in Faraday housing, as with all other ANITA electronics.

The telemetry link served two vital purposes during operation. The first was a data-linkup, with line-of-sight (LoS) transmission used when available and Tracking and Data Relay Satellite System (TDRSS) at other times. This data-link allowed for monitoring of instrument health (temperatures, battery health etc), as well as providing the ANITA collaboration with access to a small sample of data shortly after it was taken. The data linkup bandwidth was significantly smaller than ANITA's data-acquisition rate, with LoS and TDRSS providing maximum rates of 300 kbs and 6 kbs respectively (a compressed and stored ANITA-2 event is ∼30 kB and trigger rates during flight were typically 10 Hz). It was therefore impossible to transmit more than a small percentage of data, a prioritisation system ensured that the most scientifically interesting events, along with a small minimum bias sample, were relayed to ground. In the event of a failure in flight that rendered the payload unretrievable, the prioritisation system should have ensured that any neutrino candidates were already stored on the ground (see Sect. 4.5).

The second service provided to science missions by the CSBF telemetry package is the ability to send basic commands to the payload during flight (via LoS, TDRSS and IRIDIUM satellite systems). This commanding was used, for example, to provide or deny power to certain system and in the setting of trigger thresholds. Aside from periods when the payload was within LoS, the availability of commanding was limited to brief periods once per hour.

4.2.5 Position and Orientation Information

A mission critical requirement of the ANITA experiment was the access to accurate payload position and orientation information for each event taken during flight. For data analysis to successfully distinguish isolated (candidate physics) events from anthropogenic events, the orientation data must be accurate to sub-degree levels and ideally have no effect on the overall angular resolution of the experiment. ANITA-2 tackled this issue by carrying a suite of GPS antennas, with redundant back-up systems also flown.

Two square GPS antenna arrays, each consisting of four Magellan ADU5 GPS antennas, were situated at the top of the gondola. These provided heading information accurate to $<0.2^\circ$ and pitch and roll information accurate to $<0.5^\circ$ [18] (in practice, static pitch and roll values were used). A Thales G12 GPS antenna was also flown, also placed at the top of the gondola. The CSBF support package contained further GPS systems, with their data stored in the SIP and not recorded by ANITA-2. Four sun-sensors, a magnetometer and an accelerometer on the main ANITA-2 deck, with a further accelerometer at the top of the payload, provided back-up to GPS systems. Data from these systems and the CSBF data proved useful, as ANITA-2 experienced a loss of GPS information for $\sim 5\,\%$ of its flight.

Further to providing position and orientation information, both the G12 and ADU5 GPS systems were used for pulse-per-second (PPS) triggering of ANITA-2. The PPS signals were used to synchronise triggers with signals from two calibration transmission systems based at Williams Field (see Sect. 4.6.1). PPS was also used to generate a minimum bias data sample. The G12 antenna also provided timing information for the ANITA-2 flight computer's network time protocol (NTP) internal clock.

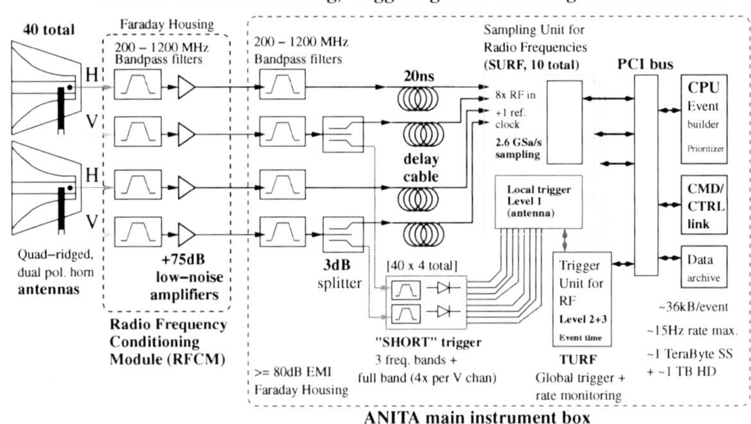

Fig. 4.6 The ANITA-2 signal chain. Figure from [19]

4.3 Radio Frequency Signal Chain

Radio frequency (RF) signals observed by each of ANITA-2's eighty channels were filtered, amplified and digitised before being stored on board during flight as uncalibrated ADC counts. A summary of the RF signal path, from front-end antennas through to digitisation and storage is shown in Fig. 4.6. Electric fields at the payload $O(0.1)$ mV/m were detectable by ANITA-2, as a result of efforts to minimise thermal noise introduction in the signal chain.

4.3.1 Front End Antennas

ANITA-2 used 40 custom designed Seavey Engineering Inc. quad ridge horn antennas. The antennas were arranged over three physical 'rings': upper (two sub-layers, eight antennas in each), lower (16 antennas in one layer) and nadir (eight antennas in one layer). The face of each antenna was 0.8 m and two perpendicular feeds at the base of the antennas allowed them to operate in two linear polarisations, with ANITA-2 recording data from all antennas in both horizontal and vertical polarisations (H- and VPOL), resulting in 80 RF channels in total.

The antennas were highly directional, with an on-axis gain of 10dBi, allowing for reconstruction of observed signal directions. To maximise efficiency on neutrino signals and reject as much thermal noise as possible, ANITA's triggering systems had to operate in tight timing coincidence windows ($O(10)$ ns, see Sect. 4.4). The antennas had an impulse response, shown in Fig. 4.7, that managed to retain much of the power from impulsive signals within the first few ns [20].

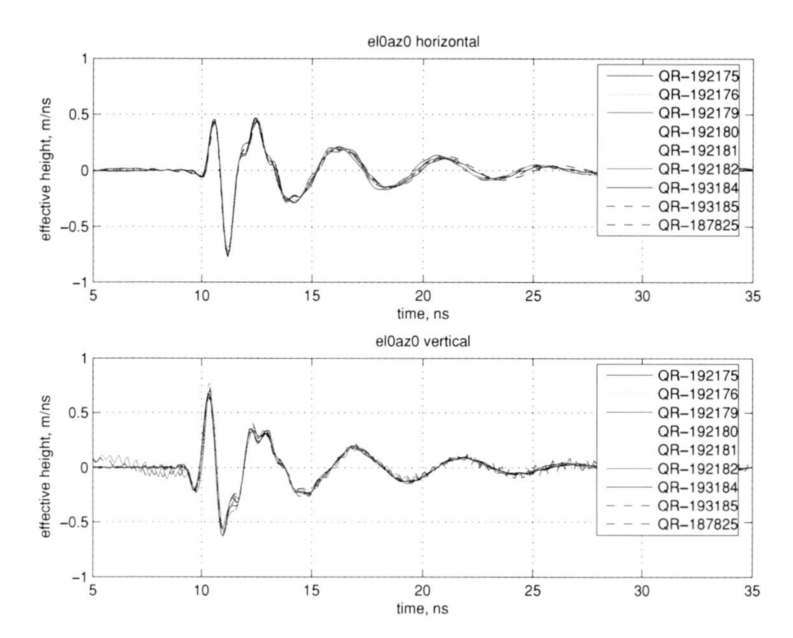

Fig. 4.7 Antenna impulse response for HPOL (*top*) and VPOL (*bottom*) for nine ANITA antennas. Plots from [2]

Figure 4.8 shows that the antennas operated with a fairly flat response over a broad frequency range, with a 3dB bandwidth of 200–1200 MHz. The antennas had a beam width of 30° at the 3 dB point, as shown in Fig. 4.9. Each antenna ring had full 360° azimuthal coverage and was split into sixteen azimuthal 'ϕ-sectors'. The upper and lower rings contained sixteen antennas each, one antenna in every ϕ-sector, the resulting 22.5° spacing providing redundant azimuthal coverage. The nadir ring contained only eight antennas, with an azimuthal spacing of 45°. Antennas were arranged with a downward cant of 10°, placing the majority of viewable ice within 5° zenith of antenna bore-sight direction and over 99 % of viewable ice within the antennas' 3dB point.

The three antenna rings were arranged with a ~3.7 m (~2.7 m) separation between top (bottom) sub-layer of the upper ring and the lower ring and a ~1.8 m separation between lower and nadir rings. The larger vertical baseline between antennas (compared to ~1 m horizontal antenna separation) resulted in a better elevation than azimuth pointing resolution in event analysis.

Due to size constraints on the ANITA-2 payload, it was not possible for the nadir ring of antennas to be in their flight positions during launch. To resolve this, the nadir antennas were attached to the main gondola frame via hinged aluminium struts which could be retracted prior to launch. A telemetry controlled actuator released the antennas after launch, with a locking system fixing nadir antenna positions during flight.

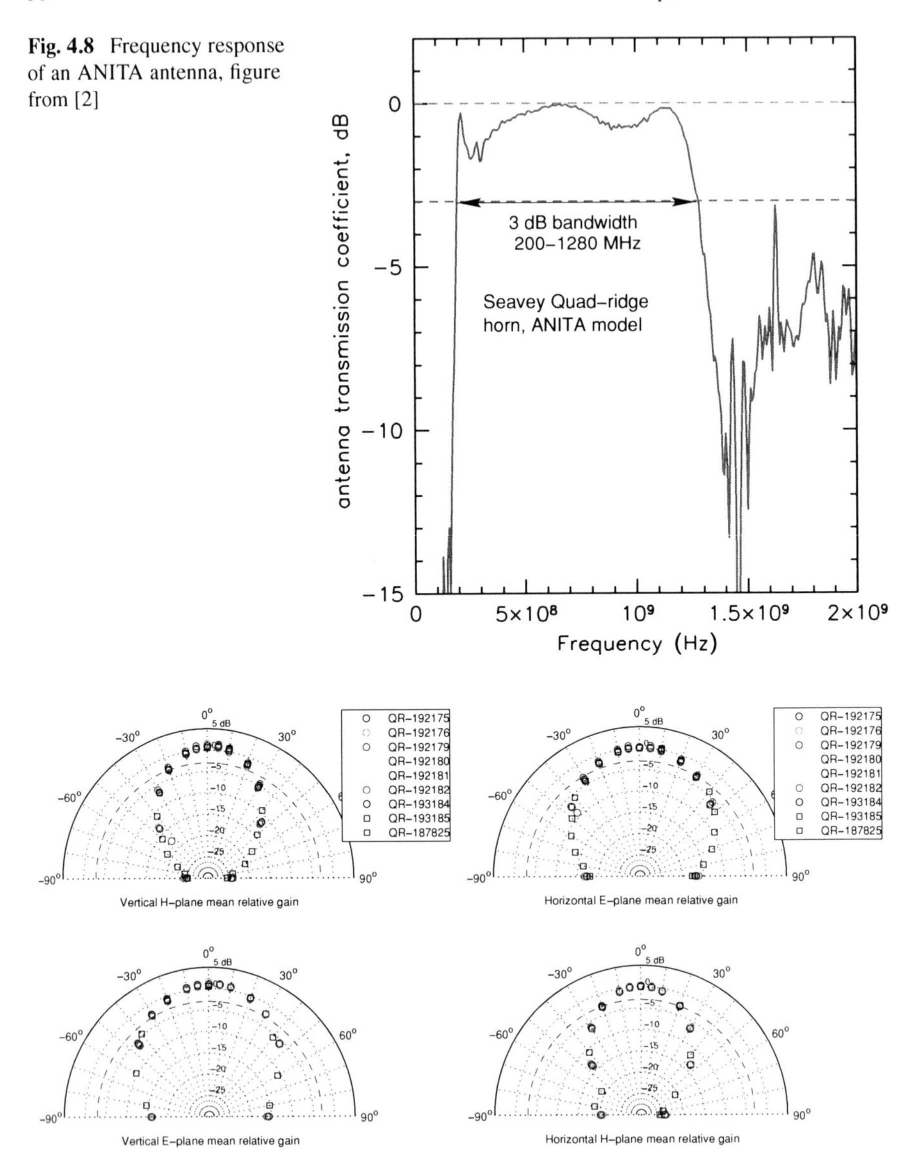

Fig. 4.8 Frequency response of an ANITA antenna, figure from [2]

Fig. 4.9 Antenna beam patterns in dB for VPOL (*left*) and HPOL (*right*) for 9 different antennas. Upper plots show response for HPOL radiation, lower plots show response for VPOL radiation. Plots from [2]

The antenna responses of Figs. 4.8, 4.7 were measured prior to the flight of ANITA-1 in an anechoic chamber [21]. These measurements were not repeated prior to the flight of ANITA-2, however, the uniformity and quality of all the antenna responses were checked by performing $S11$ (reflection) measurement using a network

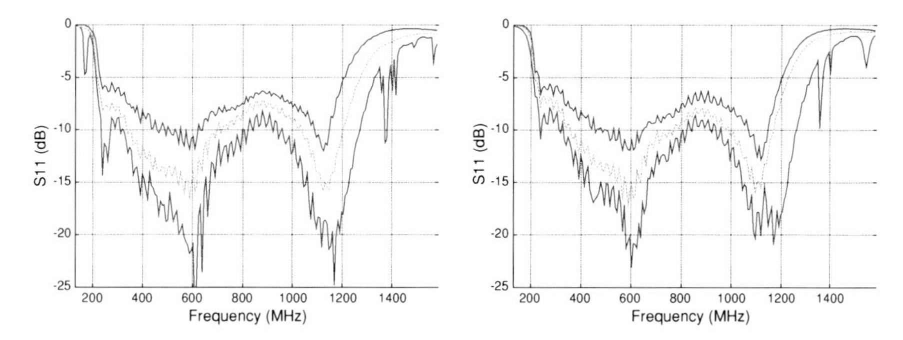

Fig. 4.10 Results of the antenna *S*11 testing for HPOL (*left*) and VPOL (*right*), central green is the average response, upper and lower red are the maximum and minimum response respectively

analyser. Figure 4.10 displays the results of these checks, with the response fairly uniform between antennas.

4.3.2 Analogue Processing

Signals received by the ANITA-2 antennas were amplified and filtered close to the antenna output. By amplifying signals close to the antenna, thermal noise contamination and transfer losses in cables are vastly reduced.

Amplification, via low-noise amplifiers (LNAs) and 200–1200 MHz filtering of HPOL channels took place in Radio Frequency Conditioning Modules (RFCMS). The RFCMs contained all modules in Faraday housing to ensure no exterior interference and were situated close to the antennas, with ~1 m of LMR-600 coaxial cable between antenna output and RFCM.

For VPOL channels, LNAs were situated in separate Faraday housing, connected directly onto the antenna outputs, these are the Antenna-Mounted Pre-Amplifier (AMPA) units. The output of the VPOL AMPAs were led, via ~1 m of LMR-600 coaxial cable into RFCMs, with each RFCM processing two VPOL and two HPOL channels.

In ANITA-1, LNAs and filtering units for all channels were housed in the RFCMs. By removing the short stretch of cable between antenna output and amplification in VPOL channels, the system noise was reduced by ~20 %. As neutrino signals are expected to be predominantly VPOL, and all triggering was conducted using the VPOL channels only, this provided a significant improvement to the hardware trigger sensitivity to neutrino-like signals.

After the front-end amplification and filtering, signals from all channels passed through a number of metres of LMR-600 cable (with length dependent on antenna location) into the main ANITA-2 instrument box.

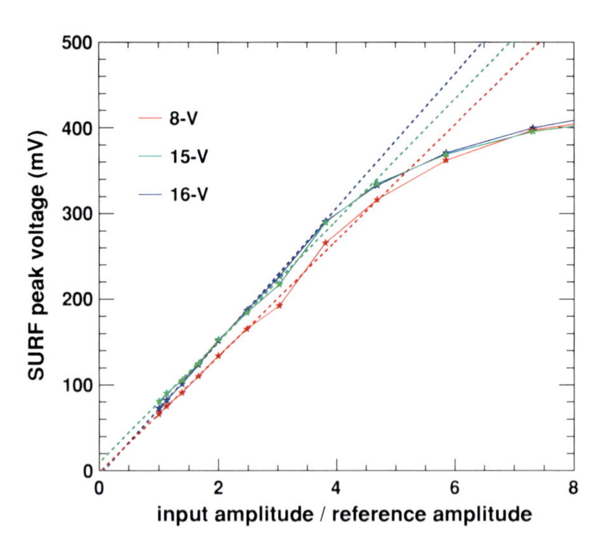

Fig. 4.11 The linearity of the amplifier response demonstrated with output pulse amplitude (taken at the instrument output) as a function of signal amplitude injected at the antenna inputs. Solid lines indicate the average amplifier response, dashed lines are linear fits to data with input/reference amplitude < 2.5. The non-linear region is almost entirely due to gain suppression from the amplifier

Effect of Analogue Processing on Signals

The amplifiers used on ANITA-2 were known to have a response which would begin to saturate for a significant fraction of events recorded in flight. Figure 4.11 illustrates that the amplifiers have a fairly linear response up to an output of $\sim$300 mV, with the response non-linear afterwards. This demonstrates that the amplifiers begin to saturate well before the maximum voltage that ANITA-2's data acquisition units were able to record ($\sim$1 V).

The gains and noise of the entire signal chain, from antenna output through to instrument box bulkhead, are shown in Figs. 4.12 and 4.13. The impulse response of signal path through to instrument output is shown in Fig. 4.14. Note the differences between VPOL and HPOL channels, caused by the difference in location of the front end LNAs, and, for Fig. 4.14, the fact that the VPOL channels are split before data acquisition.

4.3.3 Data Acquisition

Aside from the first stage filtering and amplification described in Sect. 4.3.2, all signal processing and acquisition took place in the ANITA-2 instrument box, situated on the deck above the lower ring of antennas in Faraday housing.

All data had to be stored on board ANITA-2, placing limits on the amount of data that could be stored. Meanwhile, the total power budget of an LDB flight is $<$1 kW. These combined factors meant that ANITA could not continuously digitise and record data (which would have consumed $O(10)$ W per channel, almost the entire power budget). Instead, a data acquisition architecture was developed prior

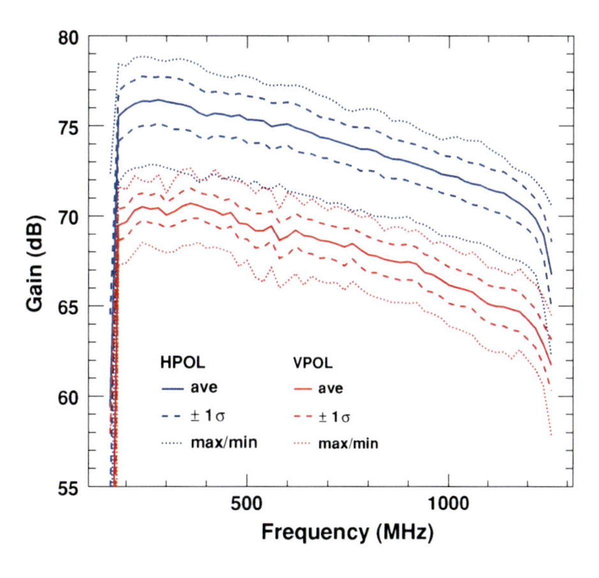

Fig. 4.12 ANITA-2 signal chain gains for HPOL and VPOL from antenna output to instrument box bulkhead

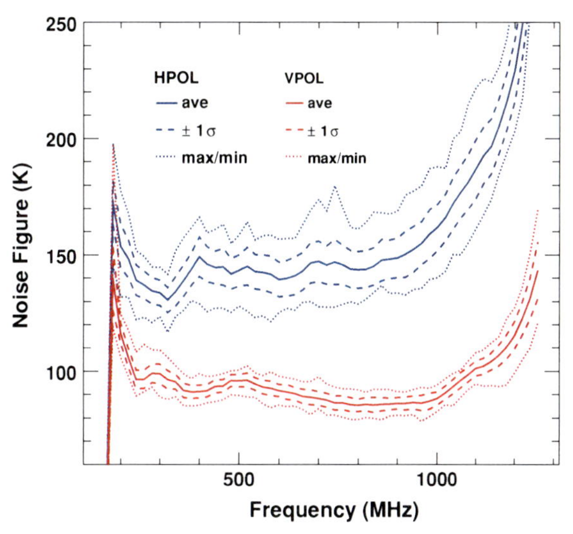

Fig. 4.13 ANITA-2 signal chain noise for HPOL and VPOL from antenna output to instrument box bulkhead

to ANITA-1, known as the LABRADOR chip, whereby ANITA could continuously sample waveform data but only digitise upon the issue of a trigger command [17]. This consumed $O(1)$ W per channel and included a four-deep buffer to minimise experiment dead-time.

Upon reaching the ANITA instrument box, signals underwent a second stage of bandpass filtering (200–1200 MHz). VPOL signals were then passed through a 3 dB splitter, with one output becoming the trigger path (Sect. 4.4) while the other was connected to the data acquisition units (DAQ). As triggering was entirely VPOL

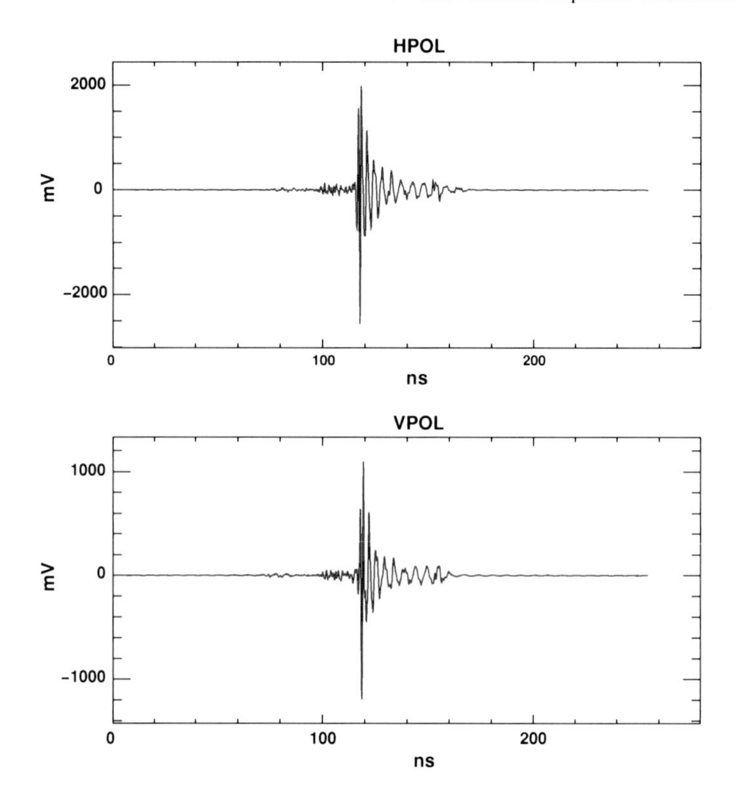

Fig. 4.14 Averaged signal chain impulse response for HPOL (*top*) and VPOL (*bottom*) channels in ANITA-2 (this does not include antenna response)

based, no splitting was applied to the HPOL channels, with all HPOL paths leading to the DAQ.

In the DAQ signal path, the signal was fed into Sampling Unit for Radio Frequency (SURF) boards. Each SURF allowed for eight input RF channels (four HPOL and four VPOL). Ten SURF boards in total provided data acquisition for all eighty RF channels, with no two adjacent antenna signals fed into the same SURF to reduce cable cross-talk between physically close channels (although signals from upper and lower antennas in the same ϕ-sector were processed on one SURF).

The signals for each channel were split and fed into four parallel LABRADOR chips, mounted on the SURF. In addition to the eight RF channels, the LABRADOR had a ninth channel for a common 125 MHz clock.

The LABRADOR chips contained a two hundred and sixty element switched capacitor array (SCA) for each of the nine input channels. The SCAs on all LABRADOR chips continuously sampled waveform data at a rate of ~2.6 Gsa/s until the issue of a trigger command, at which time sampling on the one set of SCAs was frozen and read out into a digitizer. This involved reading out analogue data corresponding to waveform amplitude and converting to digital information using nine bits of dynamic range. As each channel was sampled by four LABRADOR chips (a buffer depth of four), in the instance of the first trigger the remaining three chips would continue to sample. This significantly reduced the dead time that would be introduced by using only one LABRADOR chip for each channel, with digitisation of the frozen channels requiring $\sim$30 ms.

4.4 Triggering

The trigger system used by ANITA-2 was developed to ride thermal noise levels, allowing the experiment to be highly sensitive to neutrino signals. The trigger system used was multi-level and placed requirements on both the broadband nature and the directional coherence of observed signals. The trigger operated purely on VPOL signals, as radio emission from any neutrino interaction in the ice is expected to be predominantly vertically polarised if observed by ANITA. As described in Sect. 4.3.3, the trigger had to operate on analogue signals to circumvent power-hungry constant digitisation.

4.4.1 Level 1

After splitting, the trigger path of the VPOL channels were led into SURF High Occupancy RF Trigger (SHORT) units. Here, the VPOL signals were further split into four separate paths, with each signal path filtered into one of four bands, shown in Fig. 4.15:

- Low: 200–350 MHz
- Mid: 350–700 MHz
- High: 600–1,150 MHz
- Full: 150–1,250 MHz

The filtered sub-band channel signals were then passed through a tunnel diode, also situated in the SHORT. The tunnel diode effectively acted as a square law detector and output a unipolar pulse that tracked the power passed through the diode.

The SHORT outputs were then fed into a field-programmable gate array (FPGA), mounted on the relevant SURF board for the channel in question. Within the FPGA, signals passed through a discriminator; any signal exceeding the discriminator's threshold resulted in a positive trigger logic output which remained high for $\sim$10 ns. Importantly, the FPGA thresholds were variable, allowing them to ride the thermal noise floor levels. The thresholds were set using a servo loop, operated by the flight software (see Sect. 4.5), that regularly compared user-defined sub-band target trigger rates with the actual trigger rates in each sub-band. The software would then adjust the trigger thresholds accordingly. An example of the trigger thresholds riding thermal

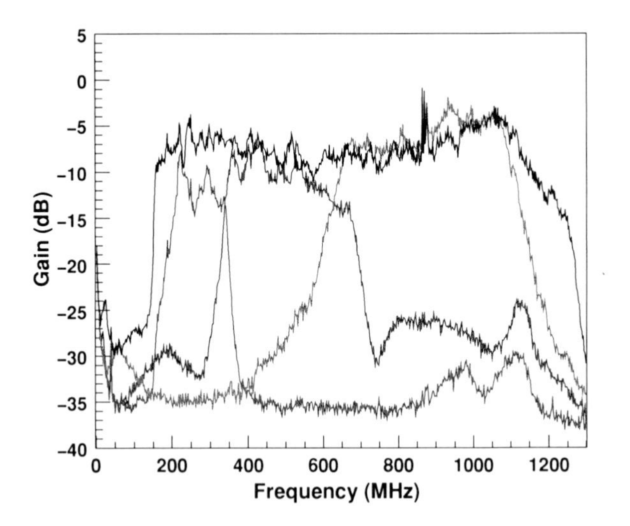

Fig. 4.15 SHORT gains as a function of frequency for each of ANITA-2's four trigger bands. The feature at ~900 MHz is likely narrowband noise in the testing environment

noise levels is given in Fig. 4.16. Typical sub-band trigger rates for antennas in the upper and lower rings were 10 MHz, 10 MHz, 6 MHz, 1 MHz for the low, mid, high and full band thresholds respectively. Nadir ring channels had typical thresholds of 200 kHz, 200 kHz, 200 kHz, 40 kHz for the low, mid, high and full band thresholds respectively.

The level 1 (L1) trigger, still within the SURF mounted FPGA, can be thought of as an antenna wide trigger. Any signal triggering 2 of 3 sub-bands (low, mid, high) as well as the full band within a 10 ns window in a single channel caused an L1 trigger. This requirement of multiple sub-bands *and* the full band being above threshold meant that there must be some broadband aspect to the incident radiation (or thermal fluctuation) for it to have passed the L1 trigger. The full band was not present in ANITA-1, which used four sub-bands (low, mid1, mid2 and high) in two circular polarisations (eight bands total), a 3 of 8 sub-band condition met the L1 requirement.

4.4.2 Level 2

The SURF based FPGAs constantly passed out trigger logic high or low output to the Trigger Unit for Radio Frequencies (TURF). This unit housed level 2 (L2) and level 3 (L3) triggering for all channels within another FPGA. On detecting an L1 trigger, the SURF based FPGA output immediately changed to high, this trigger output remained high for ~10 ns.

The L2 was an antenna ring based trigger. For either upper or lower antenna rings to trigger at L2, two of three adjacent channels on a given ring must have passed L1 trigger conditions within a ~10 ns window. The resulting L2 trigger was issued

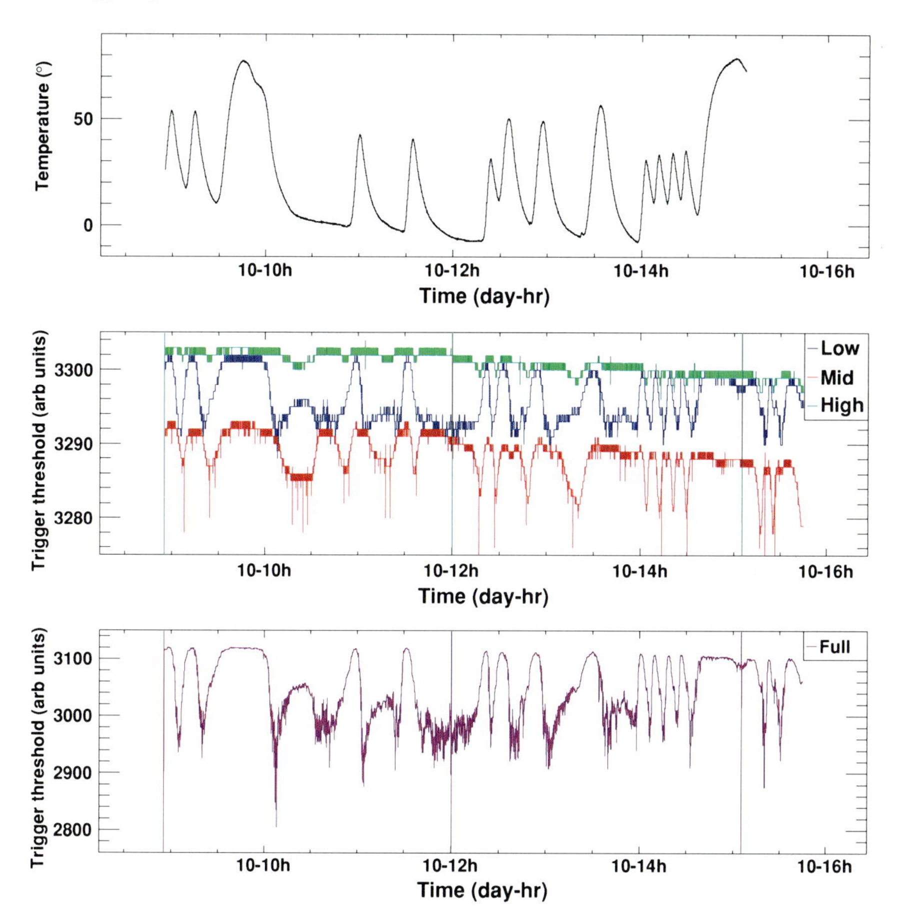

Fig. 4.16 Temperature in ϕ-sector 9 (measured at the PV array) and trigger thresholds for channel 4V (also in ϕ-sector 9) as a function of time. The trigger thresholds increase in periods when the channel experiences higher temperatures to maintain steady trigger rates. Anomalous trigger thresholds at 10-09h, 10-12h and 10-15h are the result of new data runs beginning

for the ϕ-sector corresponding to the central of the three antennas (see Fig. 4.17). For nadir antennas, which only occupied every other ϕ-sector, a channel passing L1 conditions automatically passed L2, which is why the sub-band trigger rate targets were set at lower levels for nadir channels.

4.4.3 Level 3

The L3, or global, trigger was the final trigger stage. Upon passing L3, an event was digitized and stored on board ANITA-2. This trigger stage required a L2 trigger

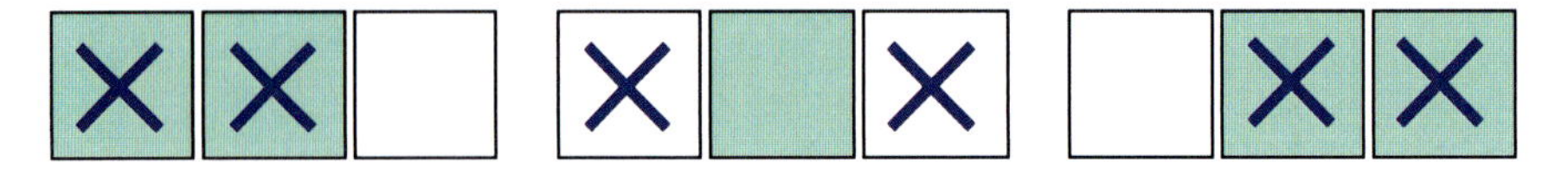

Fig. 4.17 Schematics showing three possible combinations of L1 triggers leading to an L2 trigger in the upper and lower rings of antennas. Blue crosses indicate L1 triggers, green shaded regions indicate L2 triggers. Note that only three antennas (from a ring of 16) are shown, all other antennas are assumed to have not been triggered

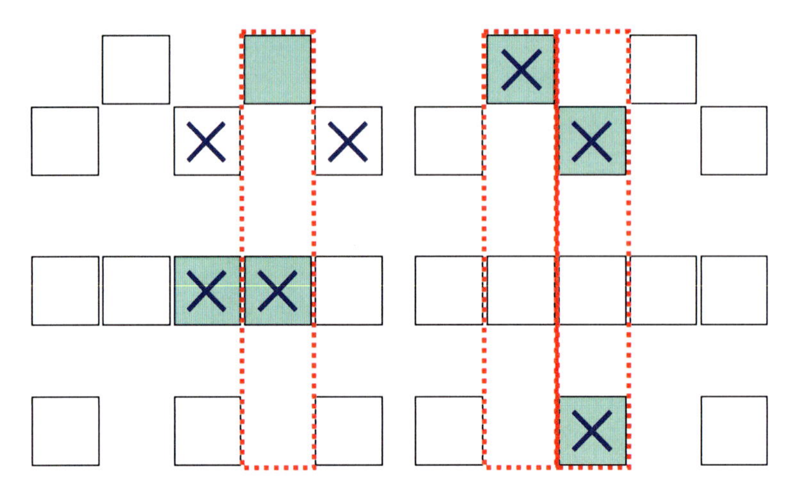

Fig. 4.18 Schematics showing two possible combinations of L1 and L2 triggers leading to a payload (L3) trigger. Blue crosses indicate L1 triggers, green shaded regions indicate L2 triggers, red dashes indicate the ϕ-sector for which an L3 trigger is assigned. Note that an L1 trigger in a nadir antenna automatically results in an L2 trigger for that sector and the two neighbouring sectors. Note that only five ϕ-sectors of antennas (from a total of 16) are shown, all other antennas are assumed to have not been triggered

command in any two of three rings within the same ϕ-sector in a $\sim$ 10 ns window (see Fig. 4.18). Both the L2 and L3 trigger requirements meant that a signal reaching these trigger stages, already required to be broadband in nature, had some geometric coincidence—as would be the case for any directional (i.e. non-thermal) event. Global trigger rates were typically 10 Hz, though rates of up to 25 Hz were observed for periods of the flight.

4.4.4 Trigger Masking

A further aspect of ANITA-2 that marked a great improvement over ANITA-1 was the introduction of directional trigger masking. The ANITA-1 flight experienced significant periods of flight time close to human bases and associated narrow-band continuous wave (CW) signals. Such emission could constantly trigger the payload, filling the buffers of the DAQ and increasing the experiment dead-time. Furthermore, these CW signals pushed sub-band trigger thresholds of channels pointing towards the CW source to very high levels, reducing instrument sensitivity. A manually operated trigger mask could be used to turn off the affected sub-band triggers; however, even with this masking, human bases caused a significant proportion of triggers when in view and reduced ANITA-1's overall sensitivity.

For ANITA-2, an automated ϕ-sector masking system, controlled by the Acqd software (see Sect. 4.5), monitored the location and rates of L3 triggers. If the software deemed that too high a rate or proportion of global triggers were originating from a specific ϕ-sector, global triggers from that sector would be masked off. This reduced the number of triggers from noisy sources and allowed thresholds from sectors of the payload unaffected by the CW source to remain sensitive to physics signals.

4.4.5 Trigger Testing

Prior to flight, the efficiency and uniformity of the L1 and L3 triggers were tested. Impulsive signals of variable strength were injected into a number of channels simultaneously, with the fraction of impulses causing a trigger recorded.

Examples of the sub-band and L1 trigger response, typical for all channels, are given in Fig. 4.19. These appear to show the low-band trigger being less efficient than the other three frequency bands. However, it should be noted that the signal being used for triggering in this instance did not pass through any signal chain, rather being injected directly into the SURF input.

For the L3 efficiency measurement, the picosecond impulse generator sent signals into the amplification units of eight VPOL channels from three adjacent ϕ-sectors. The central ϕ-sector contained an upper and lower channel. The two outer ϕ-sectors contained upper, lower and nadir channels. An extra 3dB of attenuation was applied to the outer channels to compensate for the lack of antenna off-axis response.

The results of the trigger testing are shown in Fig. 4.20. It was found that, for the picosecond pulser used, ANITA had a hardware trigger efficiency of 50 % at in input pulse SNR of 3.23. This marked a significant improvement over ANITA-1, which displayed a 50 % hardware trigger efficiency at an input pulse SNR of $\sim$5.5 for a similar test setup [2].

The effect of removing each ring of antennas from contributing to a trigger was investigated, results are shown in Fig. 4.21. It was found that masking the upper and lower rings of antennas was more detrimental to triggering efficiency than masking the nadir ring of antennas. However, even with entire rings of antennas inactive, the payload remained sensitive to signals with SNR values <4.

This hardware efficiency must be taken with consideration of the fact that the frequency content of the pulser may differ from that of an Askaryan signal. Also, although the hardware trigger rate was approximately equal to that during flight (10 Hz when triggered by thermal events), trigger rate targets for each sub-band differed during flight.

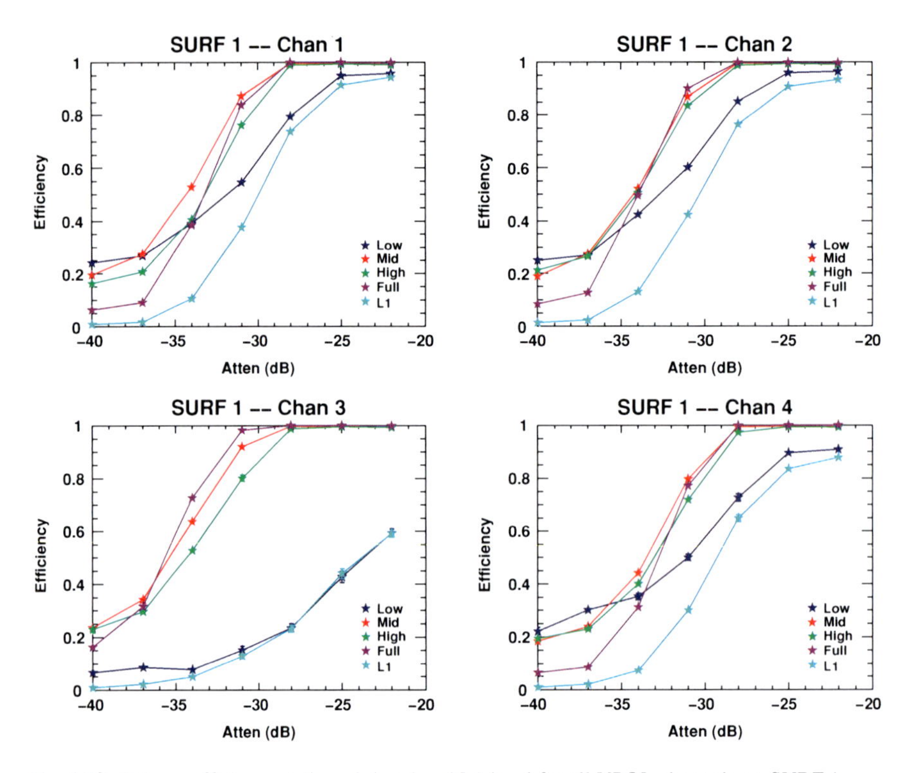

Fig. 4.19 Trigger efficiency at the sub-band and L1 level for all VPOL channels on SURF 1

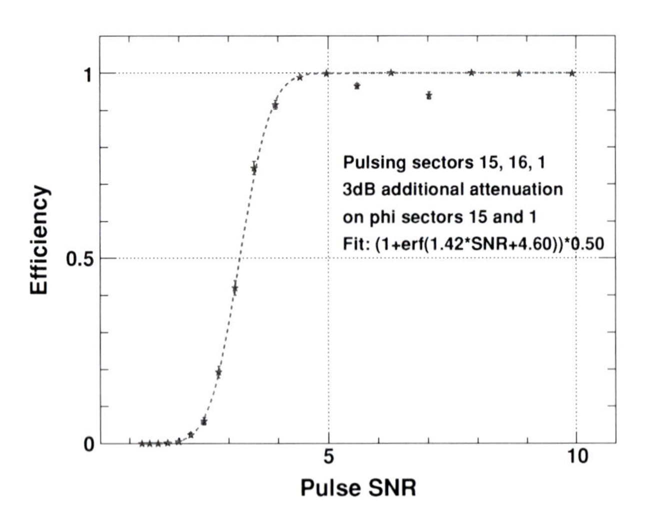

Fig. 4.20 ANITA-2 global trigger efficiency

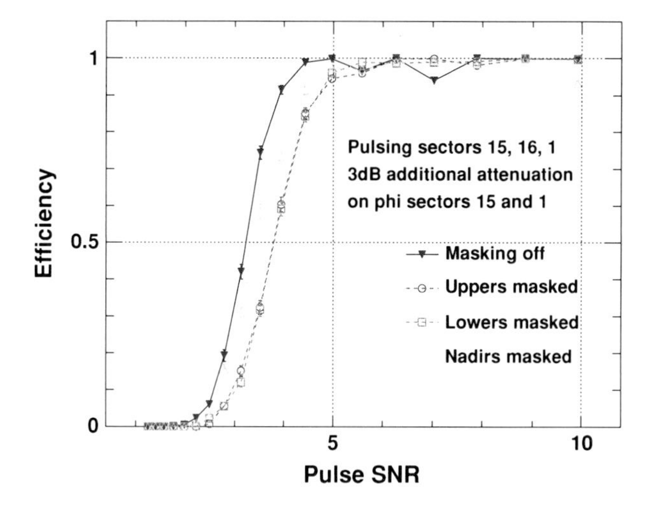

Fig. 4.21 ANITA-2 global trigger efficiency with masking of antenna rings turned on

4.5 Flight Software and Data Storage

The ANITA-2 flight computer, a cPCI single-board computer, was housed within the main instrument box. Flight software was split into a number of autonomous programmes, with the overall hierarchy shown in Fig. 4.22.

All RF data received by the ANITA-2 antennas was processed via the acquisition daemon (Acqd). This included acquiring waveform and housekeeping data from the SURFs and trigger and associated timing information via the TURFIO. As mentioned previously, Acqd was also responsible for setting the trigger thresholds and ϕ-sector masking.

After being processed via Acqd, event data was written to disk by the event daemon (Eventd), prioritizer daemon (Prioritized) and archive daemon (Archived). Prioritized additionally assigned the event a relative significance on a $1-9$ integer scale (1 being the most neutrino like). This was achieved by 'box-car' smoothing of the event (removing thermal noise fluctuations) and performing a quick cross-correlation, then comparing the locations and significance of signal peaks. Events with a sufficiently high priority were then queued and relayed to ground via LOSd or SIPd (line of sight daemon and SIP daemon respectively).

The data processed by Acqd represented 98 % of stored data. The remaining 2 % consisted of housekeeping data, with voltages and currents of all the main power subsystems, on-board temperatures and various other data. Trigger rates, GPS data and the status of disk drives were also recorded.

Three sets of data storage devices were flown on board the ANITA-2 instrument. Two arrays of eight 1TB MTRON solid state drives were used along with a spinning hard-drive that was housed in a shock mounted pressure vessel called the 'neobrick'. Each of these three storage devices held an independent set of flight data.

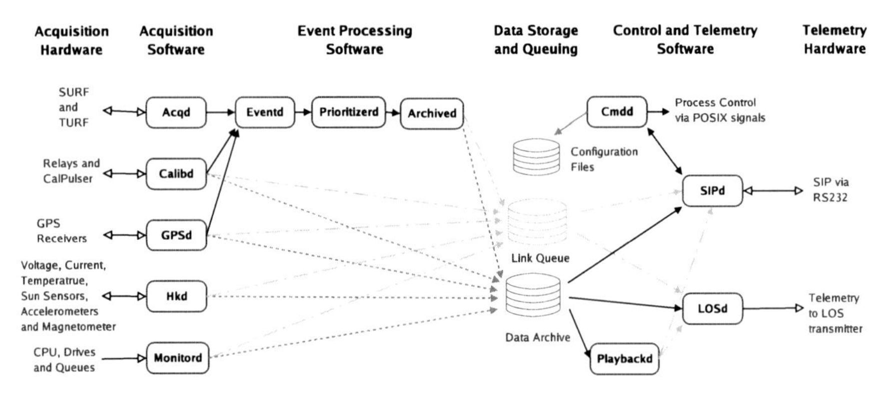

Fig. 4.22 A schematic of the ANITA-2 flight software architecture, figure from [2]

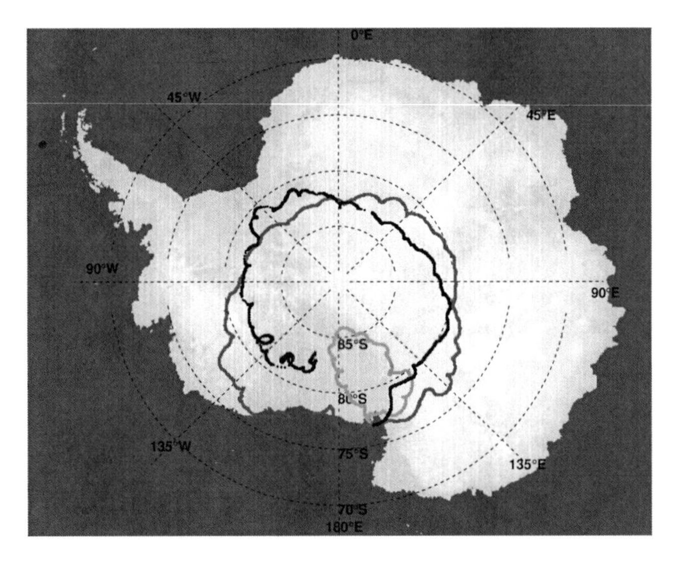

Fig. 4.23 The ANITA-2 flight path. Green, red and black markers are used for the flight path following the first, second and third pass over McMurdo station respectively

4.6 ANITA-2 Flight

ANITA-2 launched from Williams Field, close to McMurdo station, on 21st December 2008 and was aloft for a total of 31 days, with 28.5 days of live-time. The flight path, shown in Fig. 4.23, included two passes over the deep ice of east Antarctica. The average ice depth in ANITA-2's field of view was 1.5 km. The flight was terminated on 21st January 2009, with the payload landing at 83.34°S, 162.22°W, close to Siple Dome base. During flight, ANITA-2 recorded 26.7 million triggers, of which approximately 21 million were RF-induced.

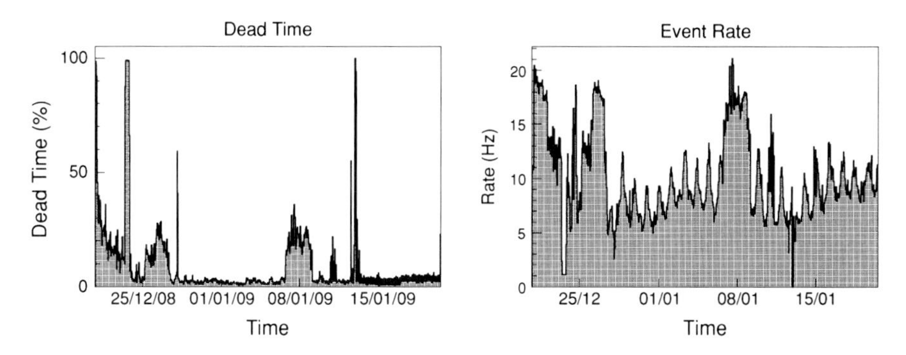

Fig. 4.24 ANITA-2 dead time and trigger rates during flight. Plots by R. Nichol

4.6.1 Ground Calibration Pulses

Ground calibration pulsing systems were used during the ANITA-2 flight to send impulsive signals to the payload. These systems were used to monitor instrument response during flight, calculate trigger efficiencies and inform analysis techniques. Calibration signals were transmitted from two locations: Taylor Dome, a field camp location approximately 200 km west of McMurdo station; and Williams Field, the balloon launch location.

The Taylor Dome site consisted of a dipole antenna transmitting impulsive signals from a 90 m deep borehole. The system was powered by PV cells and, after being set-up and tested by a team from UCLA, was left to run automatically at a rate of 1 Hz throughout the ANITA-2 flight. Signals from the Taylor Dome calibration antenna were observed by ANITA-2 on three separate passes.

At the Williams Field site, another dipole antenna was placed in a 25 m deep borehole, while a Seavey antenna identical to the flight antennas was used to transmit signals in horizontal, vertical and 45° polarisations. Unlike the Taylor Dome system, which was fully automated with events only recorded by ANITA-2 if they caused a global trigger, the Williams field systems were synchronised to ANITA-2's GPS controlled PPS triggering system. Calibration signals were sent from the Williams Field borehole on four separate passes and from the surface antenna on three passes (one of which was immediately after launch); pulsing activities ceased on 8th January 2009.

4.6.2 Performance

ANITA-2 had a successful flight when compared with ANITA-1, which suffered large amounts of dead-time due to necessary rebooting of the flight computer and high

Fig. 4.25 Photograph taken by Brian Hill shortly after arriving at the ANITA landing site

buffer occupancy when close to human bases. The fractional dead time and event rate over the course of the ANITA-2 flight are shown in Fig. 4.24. A few minor issues did result in a reduction in instrument sensitivity. Shortly after launch of the payload, one VPOL channels from the upper ring of antennas (channel 2V) failed, with an apparent lack of power reaching the channel's LNA. Although power returned intermittently, channel 2V was removed from all analysis of the flight data. Issues with the charge controller resulted in concern for ANITA-2's battery health during periods of the flight. To assuage fears that the instrument might lose power completely, power to the RFCMS (and hence, amplification) on the nadir ring of antennas was cut off for $\sim$5 % of the flight, with ANITA-2 operating with 32 antennas during such times. Loss of GPS data for $\sim$5 % of the flight necessitated the use of sun-sensor and CSBF GPS data. The sun-sensor data is less accurate than the ADU5 orientation information. Finally, the MTRON solid state drives were rendered unusable after 22 days of flight, though no data on the drives themselves was corrupted. The neobrick drive provided reliable data-storage throughout flight, resulting in no loss of data.

Figure 4.25 shows the ANITA-2 payload after the termination of its flight. A full recovery of the instrument was possible, each mission critical piece of hardware was recovered and transported off the Antarctic continent. Only a small number of antennas and associated gondola parts were not shipped from Antarctica by the end of the Austral Summer. All three sets of data, the two arrays of MTRON solid state disks and the Neobrick, were undamaged during landing.

References

1. S. Barwick, D. Besson, P. Gorham, D. Saltzberg, J. Glaciol. **51**, 231 (2005)
2. The ANITA Collaboration, P. Gorham et al., Astropart. Phys. **32**, 10 (2009), [astro-ph/0812.1920]

3. The ANITA Collaboration, P.W. Gorham et al., Phys. Rev. Lett. **103**, 051103 (2009), [astro-ph/0812.2715]
4. The ANITA Collaboration, S. Hoover et al., Phys. Rev. Lett. **105**, 151101 (2010), [astro-ph/1005.0035]
5. S. Hoover, in *A Search for Ultrahigh-Energy Neutrinos and Measurement of Cosmic Ray Radio Emission with the Antarctic Impulsive Transient Antenna*, PhD thesis, (University of California, Los Angeles, 2010)
6. The ANITA Collaboration, P.W. Gorham et al., Phys. Rev. **D82**, 022004 (2010), [astro-ph/1003.2961]
7. A. Vieregg, in *The Search for Astrophysical Ultra-High Energy Neutrinos Using Radio Detection Techniques*, PhD thesis, (University of California, Los Angeles, 2010)
8. The AMANDA Collaboration, M. Ackermann et al., Astrophys. J. **675**, 1014 (2008), [astro-ph/0711.3022]
9. The ANITA Collaboration, S. Barwick et al., Phys. Rev. Lett. **96**, 171101 (2006), [astro-ph/0512265]
10. The Pierre Auger Collaboration, J. Abraham et al., Phys. Rev. Lett. **100**, 211101 (2008), [astro-ph/0712.1909]
11. N.G. Lehtinen, P.W. Gorham, A.R. Jacobson, R.A. Roussel-Dupré, Phys. Rev. D **69**, 013008 (2004)
12. The High Resolution FlyGs Eye Collaboration, R.U. Abbasi et al., Astrophys. J. **684**, 790 (2008), [astro-ph/0803.0554]
13. The RICE Collaboration, I. Kravchenko et al., Phys. Rev. D **73**, 082002 (2006)
14. R. Gandhi, C. Quigg, M.H. Reno, I. Sarcevic, Phys. Rev. **D58**, 093009 (1998), [hep-ph/9807264]
15. A. Cooper-Sarkar, S. Sarkar, JHEP **01**, 075 (2008), [hep-ph/0710.5303]
16. T. Barrella, S. Barwick, D. Saltzberg, astro-ph/1011.0477
17. G.S. Varner et al., Nucl. Instrum. Meth. **A583**, 447 (2007), [physics/0509023]
18. Magellan Navigation, ADU5 Operation and Reference Manual. Available online from http://www.thalesnavigation.com
19. K. Belov et al., ANITA internal note, 2008
20. The ANITA Collaboration, P.W. Gorham et al., Phys. Rev. Lett. **99**, 171101 (2007), [hep-ex/0611008]
21. P. Miocinovic, ANITA internal note, 2006

Chapter 5
Event Simulation

The default UHE neutrino Monte-Carlo simulation used by ANITA is *icemc*, written by Amy Connolly and others [1]. This simulation generates neutrinos from a given energy spectrum and propagates them through the Earth. Upon a simulated neutrino interacting in ice (or other desired medium), *icemc* generates and propagates the resultant Askaryan radiation. *Icemc* also simulates the ANITA-2 payload, with the response to radio signals, a trigger simulation and thermal noise generation. Finally, *icemc* contains an array of editable options, providing control over a range of aspects from the UHE neutrino energy spectrum through to ANITA trigger banding and logic. A schematic of the steps required for a simulation that will inform us of ANITA's sensitivity to UHE neutrinos is given in Fig. 5.1.

Waveform outputs from *icemc* are not of the same format as ANITA-2 data, meaning that an assessment of analysis techniques on simulated neutrinos for ANITA-2 was not possible. Problems with low signal-to-noise (SNR) simulated signals resulting in a simulated trigger were discovered when investigating analysis sensitivity to *icemc* events (see Chap. 6).

Because of these issues, an *icemc* add-on, called *icemcEventMaker*, was developed with the help of Ryan Nichol to simulate the response of ANITA-2 to RF signals. This instrument simulation takes the electric field produced by an *icemc* neutrino and applies antenna and signal chain responses before outputting waveform data in the same format as real ANITA-2 events. A number of the features of the *icemcEventMaker* code were adapted from Stephen Hoover's ANITA-1 simulation code [2]. The simulation includes a trigger, based on the *icemc* power-integrating time-domain model [3]. *IcemcEventMaker* was constructed to replicate flight conditions (e.g. unique noise levels in each channel) such that tools used for ANITA-2 analysis can be tested and provide outputs as close to those of the real data outputs as possible.

M. J. Mottram, *A Search for Ultra-High Energy Neutrinos and Cosmic-Rays with ANITA-2*, Springer Theses, DOI: 10.1007/978-3-642-30032-5_5,

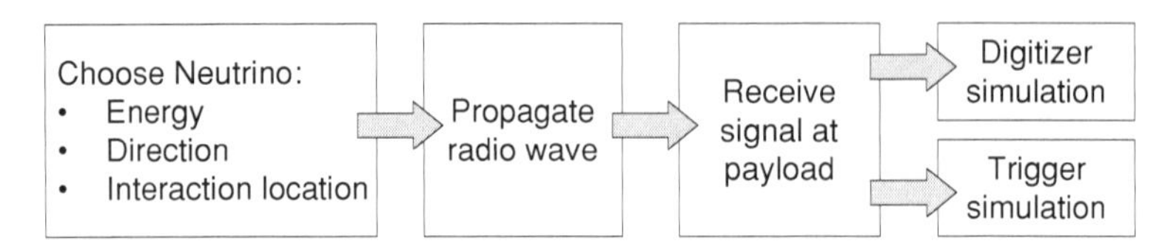

Fig. 5.1 Schematic of the processes required for ANITA UHE neutrino simulations

5.1 Approach to Instrument Response Simulation

IcemcEventMaker takes neutrino induced E-fields from *icemc* and creates ANITA-2 format output data. The responses of both the signal and trigger paths are included, allowing tests of both analysis and hardware efficiencies to neutrino signals.

5.1.1 Signal Path

For each event, the E-field at payload from *icemc* is convolved with a measured instrument response. The signal is first convolved with antenna responses (see Fig. 4.7), which accounts for the angle of incidence of a signal at each antenna. The output of the antenna is then convolved with the remaining signal chain (see Fig. 4.14). This provides waveform information as it would appear at the SURF output, after the entire signal chain. Thus far signals are purely from the expected Askaryan emission, with no noise included. Noise for each event is generated at the SURF output, as described below, and superimposed on the pure Askaryan signal. An example of a simulated Askaryan signal being received by the ANITA-2 payload is shown in Fig. 5.2.

5.1.2 Thermal Noise Generation

IcemcEventMaker generates thermal noise for every channel in every simulated event. The noise created is based on real flight thermal levels, with noise levels unique to each channel. In the frequency domain, the amplitude of thermal noise in a given frequency band at the SURF output will follow a Rayleigh distribution (assuming no noise source other than thermal):

$$\text{Rayleigh p.d.f} = \frac{A}{\sigma^2} e^{\frac{-A^2}{2\sigma^2}} \tag{5.1.1}$$

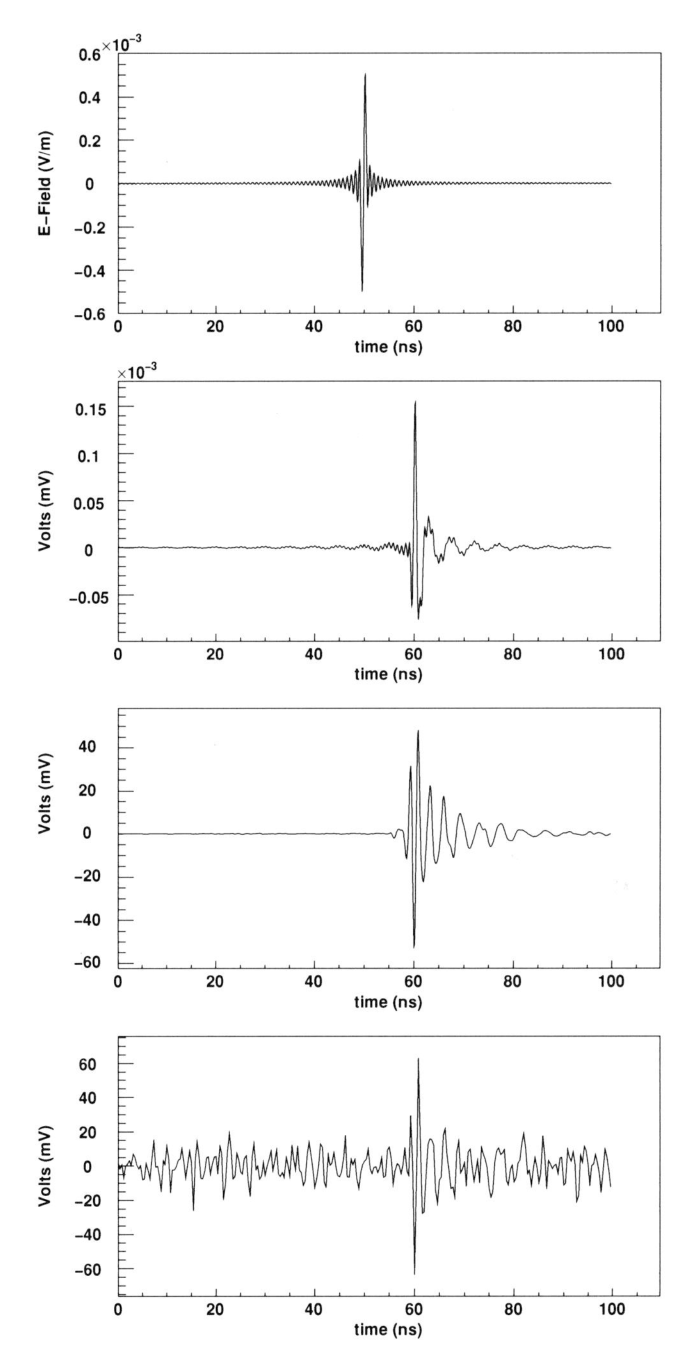

Fig. 5.2 A simulated Askaryan signal at the payload (*top*), the signal as it appears out of the antenna (*upper middle*), the signal at the SURF stage before (*lower middle* and after noise is superimposed (*bottom*). High frequency and acausal features in the *top two panels* are an artefact of the simulation and removed through filtering in the instrument output

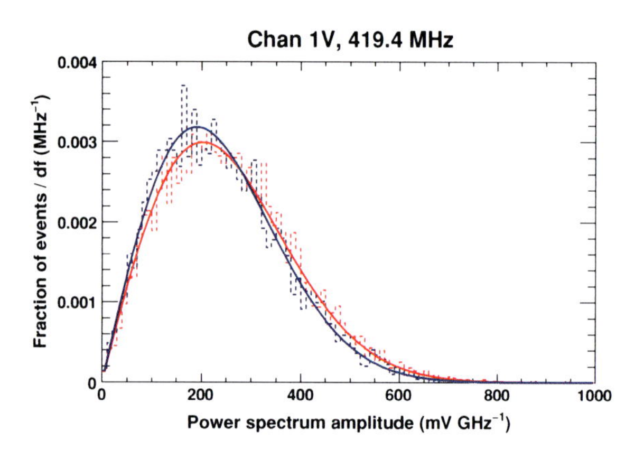

Fig. 5.3 Thermal noise amplitudes in channel 1 V at 419 MHz for real flight events within 10° of the Sun (*red*) and facing >170° from the Sun (*blue*). *Solid line* fits assume the data follow Rayleigh distributions

where A is the amplitude of the power spectrum at a frequency f. The phase of the noise, meanwhile, will be random. It is therefore possible to characterise thermal noise for a range of frequencies in each channel using real flight data.

To generate these noise distributions, data were selected from runs when ANITA-2 was far from any radio-loud human bases, runs 190–192 inclusive were used. Every event from the selected data was considered, using both RF induced and minimum bias triggers. Events failing basic quality checks (e.g. corrupt data, see Sect. 6.2.3), and any channels within or neighbouring a ϕ-sector with an L1, L2 or L3 trigger were removed from the sample.

Power spectra of the remaining events were created. Amplitude data as a function of frequency were then obtained for each channel, with Rayleigh fits created for each distribution of amplitudes. Examples of this are shown in Fig. 5.3. As channels pointing towards the Sun for a given event will have higher thermal noise levels than those that point away from the Sun, these amplitude distributions for each frequency and channel are further divided into 10° bins as a function of angle to the Sun.

Finally, as satellite noise contaminates all ANITA-2 events in a narrow (~40 MHz) range around 260 MHz, amplitudes in the 200–300 MHz range are taken by extrapolating noise data from a quiet band (350 MHz) based on amplitude vs frequency distributions from thermal data taken during ANITA-2 integration prior to flight.

Noise is generated on an event by event basis in the frequency domain, then transformed to a time domain waveform and superimposed on the neutrino signal before the event is passed through the trigger or recorded.

5.1.3 Trigger Path

As Fig. 4.6 shows, signals passed through the trigger were separated prior to the SURF from those that were recorded and stored by ANITA-2. To obtain the signal for the trigger chain, a copy of the simulated output data has the SURF gains (Fig. 5.4) removed. Gain data from the ANITA-2 SHORTs (Fig. 4.15) are then applied to one of

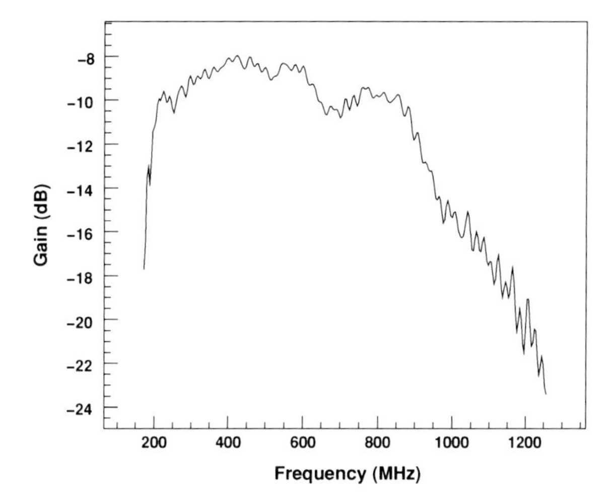

Fig. 5.4 SURF gain as a function of frequency. Simulated signals at the SURF output have the SURF gain removed before passing through the trigger chain (no phase adjustment is applied)

four copies of the signal, representing the four sub-bands used in triggering. Trigger responses for each band, taken from a model developed for *icemc* [3], are convolved with the data to provide the trigger chain outputs.

The triggering system in ANITA used tunnel diodes with variable thresholds. The diode model developed for *icemc* is based on two requirements:

- The diode output should average to zero
- The diode function should closely match the diode output when given a delta function input

Using measured diode outputs from systems testing, along with the above requirements, resulted in diode functions that can be approximated by gaussian functions followed by an exponential drop off. Trigger responses for each of the four bands are shown in Fig. 5.5.

5.1.4 Threshold Setting

As described in Sect. 4.4, ANITA-2's absolute trigger thresholds at the sub-band level, which monitor power, were variable. The thresholds can be thought of as a value relative to $\langle P \rangle$, the average power. By maintaining thresholds as a constant function of $\langle P \rangle$ they were able to ride thermal noise levels, allowing trigger rates to remain relatively constant. *IcemcEventMaker* uses a relationship between the trigger thresholds and trigger rates (trigger rates were set manually during flight) shown in Fig. 5.6.

To set trigger thresholds, *IcemcEventMaker* processes a number of noise events (typically 500) prior to running over any further data. After passing these thermal events through the trigger, a value of $\langle P \rangle$ is calculated. Taking the trigger rate targets

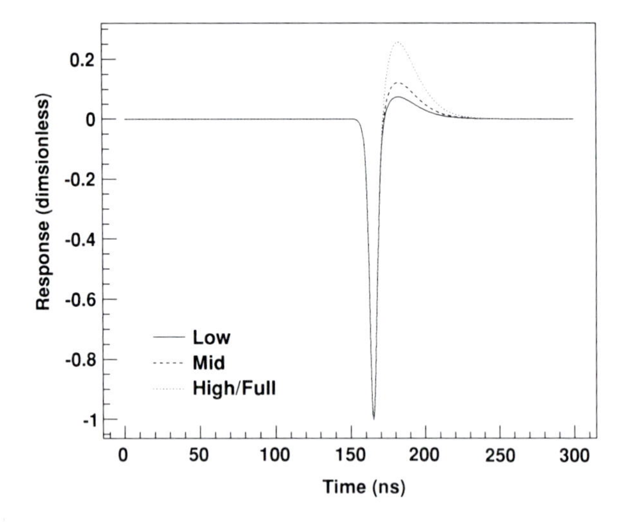

Fig. 5.5 Tunnel diode (power integrating trigger) responses for the four trigger bands from the model used in [3]

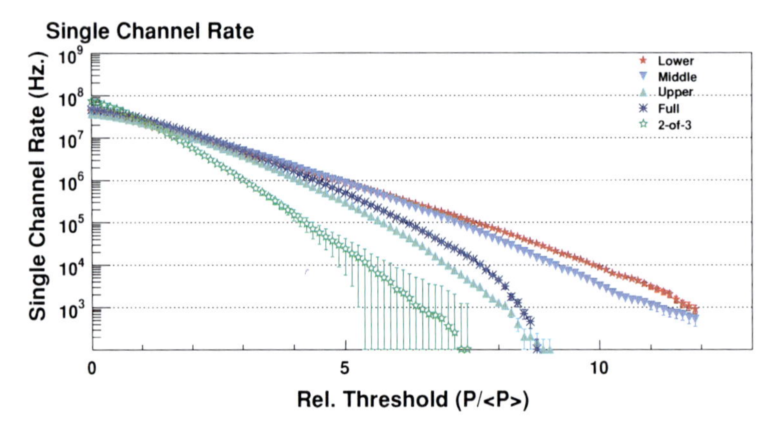

Fig. 5.6 The dependence of the ANITA-2 L0 and L1 trigger rates on the trigger thresholds. Figure by R. Nichol

used during flight, the trigger threshold for each sub-band and each channel can be set.

5.2 Simulated Hardware Efficiency

The trigger efficiency of ANITA-2 was measured prior to flight during the ANITA-2 integration at Palestine, Texas. Here, the output of a picosecond pulser was injected into the ANITA-2 signal chain between the antenna and amplification stage, as

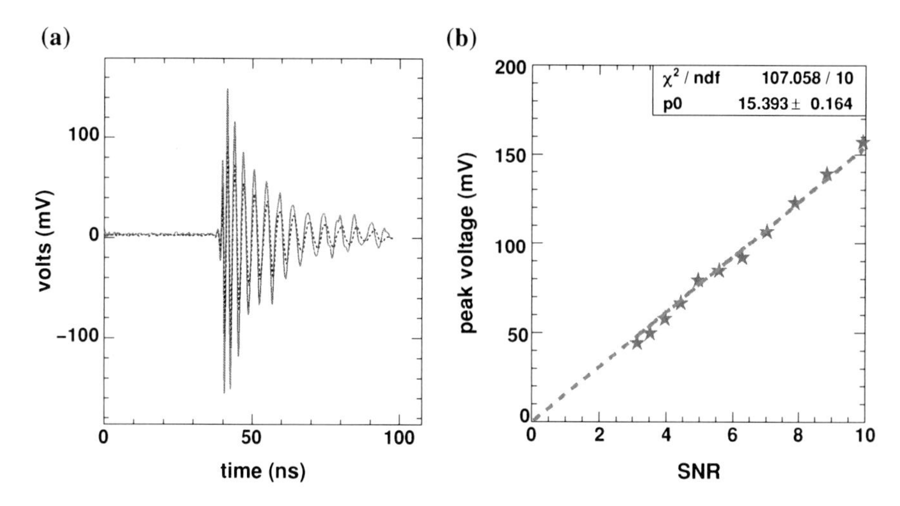

Fig. 5.7 **a** A characterised signal from the Palestine integration trigger testing, with inserted signal SNR of 9.92 in the central ϕ-sector (*red, solid*) and adjacent ϕ-sectors (*blue,dashed*). Signals are shown before noise is overlaid. **b** Extrapolation of the signal peak voltage versus SNR to signal strengths that could not be characterised. Plot shown for central ϕ-sector, lower ring, only

discussed in Sect. 4.4.5. The hardware efficiency was measured to be 50 % for an input pulse SNR of 3.23 (Fig. 4.20). The *icemcEventMaker* trigger simulation can be compared to the real trigger efficiency using characterised picosecond pulses, based on data from the ANITA-2 integration.

Noise was characterised in much the same way as the in flight simulated noise, using untriggered channel noise from runs immediately following the efficiency testing. Signals, meanwhile, were characterised by taking a number of runs when the input pulse SNR was high enough to provide a near 100 % trigger efficiency. Averaging over all the pulses in the run, noise was averaged out of the waveform and the input pulse with no noise was obtained. The pulse strength was then plotted as a function of attenuation and extrapolated to attenuation settings where trigger efficiency and input SNR were low, as shown in Fig. 5.7.

Signals used for the efficiency testing during the integration were then replicated and passed through the simulated instrument response. Running *icemcEventMaker* with trigger thresholds set by passing noise through the trigger, the simulated trigger was found to be slightly ($\sim$16 %) less efficient than the real trigger. The trigger efficiency for simulated neutrino events was also calculated, the overall efficiency for neutrinos was found to be about 2 times lower than the efficiency for the picosecond pulser due to differing frequency content of Askaryan radiation and the picosecond pulser.

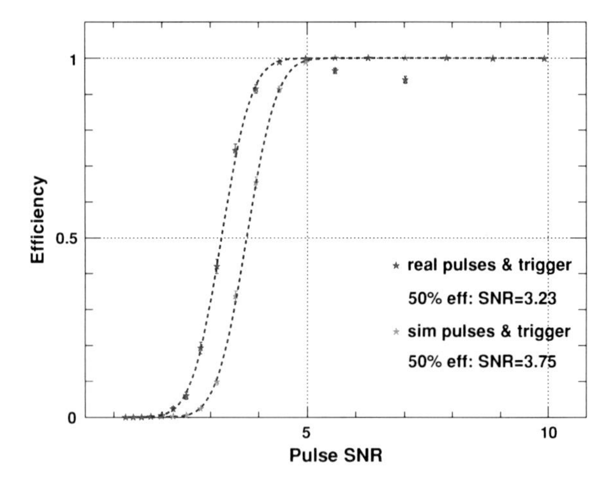

Fig. 5.8 The simulated and real efficiencies on the pulser used in the Palestine ANITA integration. The two real efficiency data points well below the fit efficiency expectation were likely caused by problems latching the trigger bits in timing during the test data runs

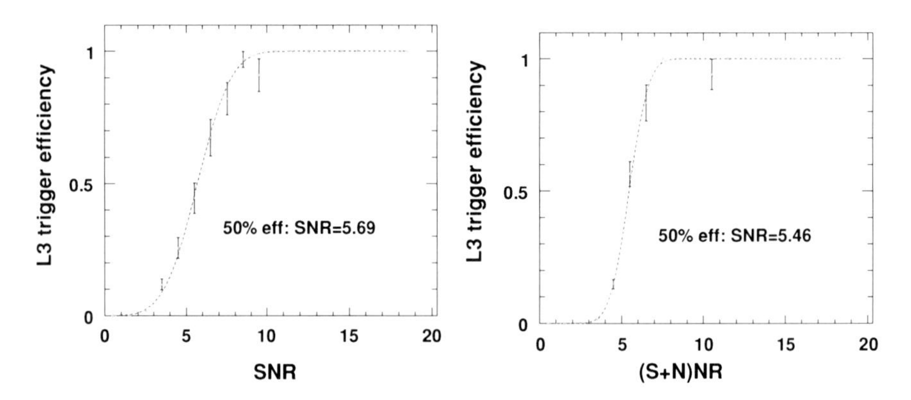

Fig. 5.9 Efficiency of the simulated hardware trigger on neutrinos vs. waveform SNR at the SURF in the facing upper antenna. True SNR—i.e. peak-to-peak of pre-noise waveform to rms of added noise is shown on the *left*, SNR after addition of noise (actually (S+N)NR) is on the *right*

5.2.1 Implications for Exposure

The reduced trigger efficiency on Askaryan-like pulses when compared to the picosecond pulser used in the Palestine integration suggests an over-optimistic trigger efficiency having been taken from the integration. By passing neutrinos of set energy through the *icemc* and *icemcEventMaker* simulation code, the effective area of ANITA-2 can be calculated (Figs. 5.8 and 5.9).

The effective volume of the ANITA-2 experiment is calculated via Eq. 5.2.1.

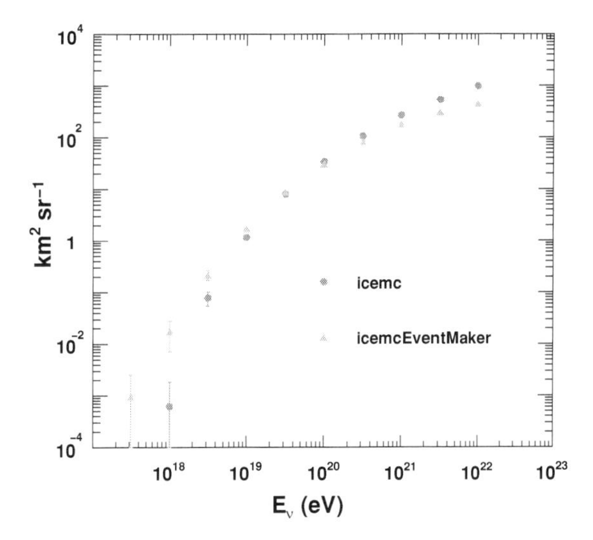

Fig. 5.10 The aperture of ANITA-2 as a function of energy with results from the *icemc* (*red*) and *icemcEventMaker* (*blue*) codes

$$V_{\mathrm{eff}}\Omega = 4\pi V_0 \frac{N_{\mathrm{det}}}{N_{\mathrm{int}}} \tag{5.2.1}$$

Here, V_{eff} is the effective volume, Ω is solid angle, V_0 is the volume of Antarctic Ice, N_{det} is the number of neutrinos detected and N_{int} is the number of neutrinos which interact. From this, an effective area can also be defined as:

$$A_{\mathrm{eff}}\Omega = \frac{V_{\mathrm{eff}}\Omega}{L_{\mathrm{int}}} \tag{5.2.2}$$

where L_{int} is the energy dependent interaction length of the neutrino. Values for this are calculated using neutrino cross sections given in [4].

The results of this unofficial exposure calculation are shown in Fig. 5.10. The *icemcEventMaker* code has a reduced exposure compared to icemc at energies $E_\nu \geq 10^{20}$ eV, meanwhile the *icemcEventMaker* has an enhanced exposure compared to *icemc* at energies $E_\nu \leq 10^{19.5}$ eV. The frequency content of the E-field at the payload is dependent on neutrino energy. At higher energies, ANITA-2 is able to observe emission from deeper interactions as well as emission from interactions where ANITA-2 is situated further off the Cherenkov cone. Both of these effects result in a reduced high-frequency cut-off in the E-field at the payload. As the instrument response differs between *icemc* and *icemcEventMaker*, this change in frequency content with energy results in the difference between exposure from the two ANITA-2 simulations being dependent on neutrino energy.

5.3 Summary

This chapter has demonstrated the development and use of an instrument simulation code, *icemcEventMaker*, for use with a Monte Carlo, *icemc*, which generates signals from UHE neutrino interactions in the Antarctic ice sheet. The instrument simulation developed underestimates ANITA's hardware efficiency by approximately 16 %, possibly due to the estimated, rather than known, tunnel diode response used within the simulated trigger.

A trigger efficiency and experiment exposure to UHE neutrinos has been calculated. This will be used in conjunction with the efficiency of any analysis to calculate any constraints on the UHE neutrino flux.

Finally, the output of the instrument simulation is in the same format as the recorded ANITA-2 data. This allows for simulated noise and simulated neutrino events to be passed through analysis codes that operate on standard ANITA-2 ROOT data formats.

References

1. A. Connolly, ANITA internal note, 2004
2. S. Hoover, A search for ultrahigh-energy neutrinos and measurement of cosmic ray radio emission with the Antarctic impulsive transient antenna, Ph.D. thesis, University of California, Los Angeles, 2010
3. A. Connolly, R. Nichol, ANITA internal note, 2008
4. A. Connolly, R.S. Thorne, D. Waters, hep-ph/1102.0691

Chapter 6
ANITA-2 Data Analysis

During its flight ANITA-2 recorded ~21.4 million RF triggered events. Simulations of ANITA-2 inform us that the experiment was likely to see very few neutrino-induced signals, with typical estimates in the range $O(0.01\text{–}1)$ events, depending on the neutrino flux model used. Meanwhile, the number of cosmic-rays that triggered ANITA-2 is expected to be lower than the 16 observed by ANITA-1 due to modifications in the trigger scheme used (see Sect. 4.4).

It is clear from these figures that the vast majority of the ANITA-2 data contains background (non-physics) events. The ANITA-2 payload design and ANITA-1 analysis inform us that most of these events were triggered by thermal noise fluctuations, with the remainder being caused by anthropogenic noise. For any cosmic-ray or neutrino-induced RF signal to be detected with statistical significance, an analysis of the ANITA-2 data must reliably remove background signals to the level of one event.

6.1 Analysis Approach

The two key backgrounds in the ANITA-2 data, thermal fluctuations and anthropogenic noise, have different characteristics. As such, analysis of the ANITA-2 data is performed in two stages. First, thermally-induced triggers, which make up the bulk of the data, are removed by requiring directional coherence of the signals in multiple antennas. After applying cuts on thermal events, a dataset of purely directional events should remain. Anthropogenic events can be identified within this sample by assuming that their source locations will be associated spatially with known bases or with repeated emission from that location.

ANITA-2 recorded waveform information in both vertical and horizontal polarisations (VPOL and HPOL respectively). The two polarisations are treated entirely separately by the ANITA-2 hardware. Furthermore, the hardware trigger only acted upon VPOL signals and the signal chains for each polarisation differ, leading to

M. J. Mottram, *A Search for Ultra-High Energy Neutrinos and Cosmic-Rays with ANITA-2*, Springer Theses, DOI: 10.1007/978-3-642-30032-5_6,

different noise levels between the two. Each event is therefore analysed separately in VPOL and HPOL and the thermal cuts are developed separately for each polarisation.

6.1.1 Blinding

A blind analysis is implemented in the most simple of manners. Any signal-like events, that is, events that contain directional information that are not coincident with any known base or other event, are hidden from the analysis until confidence in the techniques used has been demonstrated. Further to this, an unknown number of weak events from the Taylor Dome calibration antenna were inserted in the data sample at random, replacing real events (from minimum bias triggers), providing us with a measure of analysis sensitivity.

The analysis methods are trained using both noise and signal-like events. For both of these event types, real and simulated data are used in analysis training. Real data is only used if it can be removed from the main analysis data with confidence that the exposure of ANITA-2 to UHECRs or neutrinos is not being reduced, for example, minimum bias events and events containing calibration signals.

6.1.2 ANITA Data and Calibration

ANITA-2 events, once transferred from the flight storage, were converted from binary data to ROOT files that are used by all ANITA-2 analyses. The converted ROOT files included event calibration, with ADC to voltage conversion and LABRADOR chip sample number to nanosecond times made using calibration data taken in McMurdo prior to launch [1]. Channel to channel timing offsets (relating to antenna position and channel timing offsets) are also included. These offsets were calculated by Simon Bevan using a minimisation on Taylor Dome calibration pulses to fit antenna boresight locations relative to the payload GPS system along with further inter-channel timing offsets not already calibrated in [1]. Photogrammetry data that I took in Antarctica prior to flight was used by Kurt Liewer to calculate antenna positions and orientations to assist with this minimisation. Event header and housekeeping data is also included in the ROOT files, which were read into the analysis code via Ryan Nichol's eventReaderRoot libraries.

6.2 Data Samples

Analysis techniques were trained to maximise selection efficiency of RF-induced events while rejecting an appropriate level of thermal noise, with a target of half an event from thermal noise being allowed to pass analysis cuts. To achieve this, the

ANITA-2 data and simulated data were subdivided into a number of data samples, with a separate, untouched, analysis sample on which the methods could be applied.

6.2.1 Noise Training Sample

Three different samples were used for training of thermal noise rejection.

First, ANITA-2 ran a 1 Hz minimum bias trigger throughout its flight, resulting in ~2.5 million events. The majority of the events recorded were largely thermal in nature, with exceptions being periods when the payload was close to human bases. The events are only 'largely thermal' as they are all contaminated with ever-present narrowband RF from satellites (see Sect. 6.3.1). These **minimum-bias events** do not include a thermal excess coincident in time between multiple channels that would have led to the triggered events that need to be rejected.

Second, as any detectable signal arising from a neutrino interaction would originate in the Antarctic ice-sheet and any cosmic-ray-induced air-shower would develop below the balloon flight altitude, it is possible to inspect any event that reconstructs during analysis in an upward direction. A sub-set of the ANITA-2 data, using only events recorded when major bases were beyond the viewable horizon, was taken and analysed. Any event that reconstructed pointing upwards from this sample was added to a set of **upward-pointing noise events**, with the final noise-like sample including approximately 2 million events in each polarisation. Again, these events include the narrowband continuous wave (CW) noise from satellites. As triggered events, they do include the coincident signal excess between channels that lead to a global payload trigger.

Finally, **simulated noise events** were generated using the simulation outlined in the previous chapter. Over 25 million events were generated and analysed, allowing for cuts to be trained on an event sample larger than the actual ANITA-2 data. These events did not include the thermal excess, coincident in time between channels, that would lead to a global trigger and were generated without the satellite noise excess present in the real data.

6.2.2 Signal Training Sample

The Taylor Dome and Williams Field ground-calibration antennas provided over 500,000 recorded events during flight, from which signal selection efficiency could be tested. The Taylor Dome borehole antenna proved by far the most useful of these due to its more isolated location. Signals from both the Williams Field Seavey antenna and borehole antenna were observed by ANITA using the GPS PPS triggers, but the strength of background noise from McMurdo caused difficulty in analysis of the events and in calculating reliable analysis efficiency levels. Further to this, there were periods when the PPS trigger and signal transmission fell out of synchronisation, causing problems in event selection. The Taylor Dome borehole antenna, meanwhile, produced signals that allowed for both tests of analysis and trigger efficiency (as no PPS synchronisation was possible), with the events recorded by ANITA-2 containing much lower levels of CW contamination.

Neutrino signals were simulated using the icemc and instrument simulation code, as described in Chap. 5. The efficiency of the simulated trigger was not as high as that of the ANITA-2 trigger, so a number of events that did not pass the simulated trigger, but had a signal to noise ratio of >0.5, were also passed through event analysis. Any simulated event causing an L2 trigger was also passed through the analysis. The simulated neutrinos were created using the ANITA-2 flight path and timing information. A number of analysis features depend on either the payload location (e.g. event filtering) or payload status (e.g. number of antennas used). By inserting simulated neutrinos into the flight path, it was possible to analyse the event while taking into account the conditions at the simulated event time. This results in a more realistic efficiency calculation than would have been possible by assuming ideal conditions for every simulated event.

6.2.3 Analysis Sample

The entire ANITA-2 dataset is considered for analysis after first removing a number of known event classes that can confidently be assumed not to contain any desired physics signal.

Calibration Events

The analysis first reject events if they are from the set of minimum bias, other forced trigger, or calibration signals. The number of events falling into each of these samples is given in Table 6.1, with event selection outlined below.

Taylor Dome event: A simple timing cut is used when the Taylor Dome borehole antenna is within, or close to, the balloon's field of view. The time of flight for a signal to travel between the Taylor Dome antenna and the balloon is calculated. An event is classed as a Taylor Dome event if the nanosecond trigger time (number of nanoseconds after the internal clock's second mark) minus the time of flight is $-40700 < \Delta t\ (\text{ns}) < -39900$. A set of events where the signals from the Taylor Dome signal generator were reflected from the antenna up and back down the borehole cable connecting the antenna to the signal generator were also observed and selected with a cut of $-39400 < \Delta t\ (\text{ns}) < -39100$.

Williams Field event: The majority of the events from the Williams Field Seavey and borehole antennas were synchronised to the forced PPS trigger. A fraction of the Williams Field calibration signals did not coincide with the PPS trigger, some of these events can be selected using a timing cut when the calibration antennas were within the payload's field of view.

Table 6.1 Summary of the number of calibration events and the remaining number of events to be used in the main ANITA-2 analysis

Event sample	Number of events	% of total
All events	26655876	100
Taylor Dome events	116151	0.44
Williams Field borehole events	384993	1.44
Williams Field Seavey events (VPOL)	8500	0.032
Williams Field Seavey events (HPOL)	5586	0.021
Williams Field Seavey events (XPOL)	162070	0.61
Forced trigger events	4589303	17.2
Calibration pulser events	9040	0.034
Analysis sample	21380233	80.2

Forced trigger event: Any event that was caused by a software trigger (the minimum bias events) or a PPS trigger (that is not a Williams Field event) is removed from the final data sample, regardless of whether the forced trigger coincides with a hardware trigger.

Calibration pulser event: ANITA-2 flew a number of bicone antennas which sent out strong RF calibration pulses. These were transmitted in an $O(100)$ ns window at 1 Hz, typically near the start of each data run. Events are removed by applying a 100 ns timing cut and removing any event within this window with either a data flag indicating that the calibration pulser relay was on, or any events that correlated highly with a preselected sample of calibration pulses.

Event Quality Cuts

Further events are rejected based on a number of quality cuts that deem whether the instrument was operating in a satisfactory condition and whether the data was recorded without corruption, these are summarised in Table 6.2.

Saturated SURF: The SURF ADCs flown on ANITA-2 would saturate for signals above $|V| \sim 1$ V. Any event with a saturated ADC value is removed from analysis.

RFCM power off: For periods of ANITA-2's flight there were concerns about the battery status. During these times, power was turned off to the RFCMs, though data acquisition continued. Events recorded during these periods are removed either by a flag in the housekeeping data or if $V_{\mathrm{RMS}} < 20$ mV in more than ten channels.

Sync slip: On a number of occasions, ANITA-2 would experience a hardware trigger, but would record waveform data corresponding to another period of time. These events are identified by the discrepancy between TURF recorded and SURF recorded time and are removed by a flag in the event header data.

Mean Offset: Waveform data recorded by ANITA-2 should have a mean voltage close to 0 V. Events with any channel recording a mean voltage of >150 mV are removed from the analysis.

Table 6.2 Summary of events failing the quality cuts

Cut description	Events cut	Events remaining	% cut
RFCM off	1641	21378592	0.0077
Saturated SURF	57364	21321228	0.27
Self triggered blast	73722	21247506	0.34
Sync slip	18156	21229350	0.085
Mean offset	252	21229098	0.0012
Short trace	5	21229093	2×10^{-5}
No GPS	1305	21227788	0.0061
PPS+RF	8201	21219587	0.038
All quality cuts	160646	21219587	0.75

Percentages relate to the fraction of events cut from the analysis sample in Table 6.1 (21380233 events total)

Short trace: Events with channels producing waveform data <240 samples long are removed.

No GPS: About 5 % of events were recorded when the GPS data was corrupted. Although sun sensor, magnetometer and accelerometer data were used to provide payload position and orientation for most of this data, there remain a few events with no such data within 30 s of event time. These events are removed from analysis.

Self-triggered blasts: The ANITA-2 data included a number of events that were triggered by some on-payload source. This class of event was the most troublesome to remove en masse through quality cuts. The majority of these events are similar to each other in appearance, with strong, broadband signals being measured by the lower and nadir antennas, while a much lower amplitude is measured in the upper antennas. A large portion of the self-triggered blasts can be removed using cut on signal to noise ratio (SNR) difference between the nadir or lower rings and the upper ring. Events are cut if the SNR in the lower or nadir rings is at least four times that of the upper ring SNR in the same ϕ-sector of antennas. The self-triggered blasts are usually collected together in time. Further events are removed by correlating a sample of template events with every ANITA-2 event. Any event correlating particularly well that fell within a minute of an event which has been removed using the SNR ratio cut is excluded from the analysis sample. Finally, as the events originate on board, the plane wave assumption used for event analysis no longer applies, meaning that the interferometric image will not observe an isolated peak even for strong self-triggered blasts. A cut from the interferometric image is also implemented—see Sect. 6.4.

6.3 Analysis Tools

This analysis uses a small toolset with which to assess whether to retain or discard each event. These tools are designed to be able to run quickly on an event-by-event basis, with tests on how coherent the event is between channels. No major assumptions are made about the signal shape at this stage, given that no UHE neutrino interactions have been observed. However, the first flight of ANITA and subsequent data-analysis have provided information about the nature of signals that we wish to reject.

6.3.1 Event Filtering

The frequency range in which ANITA operates contains a number of bands used for communications around Antarctica. Transmission in these bands was observed throughout the ANITA-2 data, particularly when a major base such as McMurdo was within the visible horizon. Figure 6.1 shows power received by ANITA-2 as a function of frequency and time during such a period. In addition to noise from bases, satellite noise, discovered to be a nuisance in previous ANITA-1 analysis [2], is observed throughout the flight. As such, the ANITA-2 data is never free of anthropogenic CW background.

The presence of CW can be problematic, often causing a triggered, impulsive event to reconstruct towards the source of underlying CW instead. In some cases CW will cause an RF-induced event to be misreconstructed completely, pointing to neither event source nor CW source. To tackle this issue an adaptive filter is implemented to remove as much CW contamination as possible, while retaining as much original waveform information as possible.

Prior to analysis of any event, a thermal noise baseline was defined using data from the ANITA-2 flight when the balloon was far from bases. Using all minimum bias triggers from three data runs (runs 190–192, corresponding to $\sim$9 h of instrument operation), distributions of noise amplitude levels were created for each channel and frequency band in a similar manner to that outlined in Sect. 5.1.1. Using Rayleigh distributions (Eq. 6.3.1), a width (σ) is fitted to the amplitudes at each frequency for each channel.

$$\text{Rayleigh c.d.f.} = 1 - e^{\frac{-x^2}{2\sigma^2}} \tag{6.3.1}$$

Two filtering stages are employed that ensure the removal of both low level, long duration background noise and stronger, brief duration noise that contaminates only a small number of events. The first filtering method averages the power in each band and channel for 10 s of flight data, up to and including the event being analysed. The second method averages power from 3 ϕ-sectors of channels for the event in question.

The filtering first takes these averaged powers as a function of frequency ($\langle P_f \rangle$) and scales them so that the 200–1200 MHz integrated power matches that of the thermal distributions:

$$\langle P_{f,scaled} \rangle = \langle P_f \rangle \frac{\sum_{f \geq 200\,\text{MHz}}^{f \leq 1200\,\text{MHz}} P_{f,thermal}}{\sum_{f \geq 200\,\text{MHz}}^{f \leq 1200\,\text{MHz}} \langle P_f \rangle} \tag{6.3.2}$$

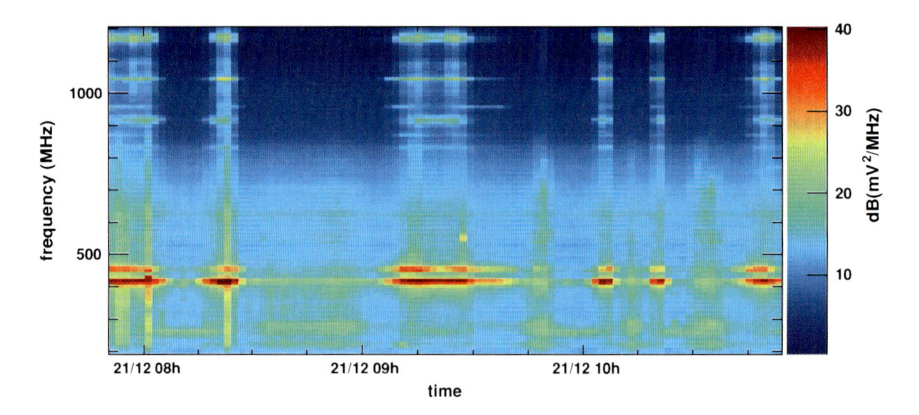

Fig. 6.1 Average power received by channel 1 V as a function of time and frequency during a 3 h period when the payload was close to McMurdo. The antenna sees CW from McMurdo in a number of bands, the strength as a function of time is caused by rotation of the payload, bringing the base in and out of view

To test whether a channel requires filtering in a given frequency band, the scaled power in this band ($\langle P_{f,scaled} \rangle$) is compared to the thermal power ($P_{f,thermal}$):

$$f_P = \exp\left(\frac{-\langle P_{f,ch,10s,scaled} \rangle^2}{2 \times P^2_{f,ch,thermal}} \right) \tag{6.3.3}$$

where f_P is some measure of the probability that observed emission in a given band is thermal or broadband in nature. If f_P falls below a preset level, then the frequency band contains too much power and is filtered.

Using an optimisation on Taylor Dome calibration signals and minimum-bias events, specific details of the filtering algorithm were chosen. Thresholds were set for the analysis of $f_P < 0.1$ for the 10 s averaged filter and $f_P < 0.02$ for the ϕ-sector averaged filter. The filtering algorithm will also filter neighbouring frequency bands until a local maximum of f_P is found, or f_P rises to twice the filter threshold. If any frequency band of a given channel is filtered, the filter is also applied to channels within the same and neighbouring ϕ-sectors.

In addition to the filtering described above, a 200–1200 MHz bandpass filter is applied to each recorded waveform, in line with ANITA-2's hardware filters. Filtered bands within the 200–1200 MHz range are then whitened. For this, scaled thermal noise amplitudes with a random phase are inserted into each filtered band. An example of filtering is shown for an event recorded when the payload was within view of McMurdo is shown in Fig. 6.2.

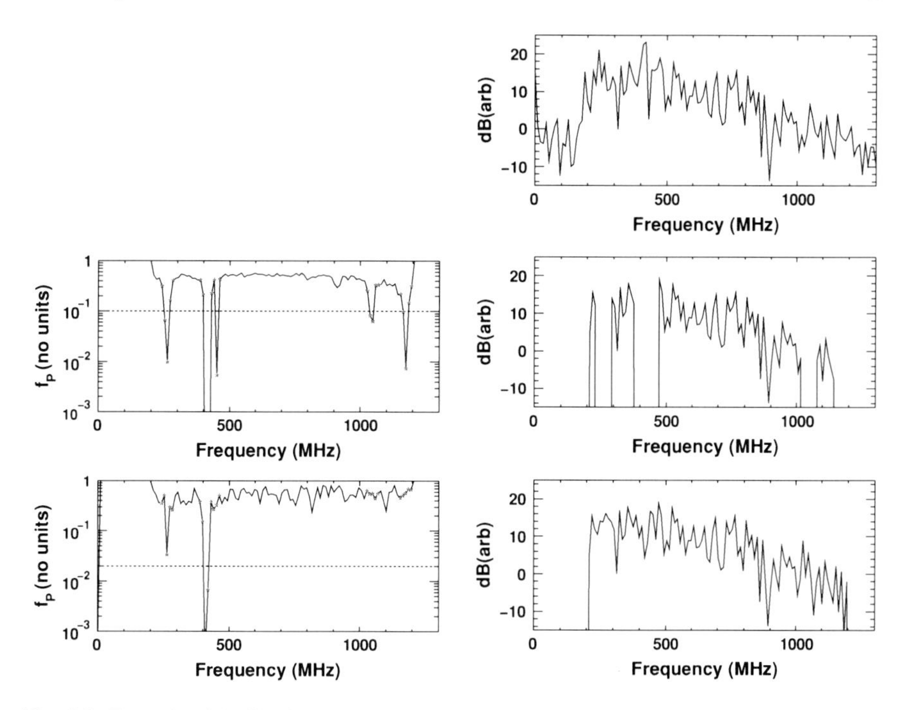

Fig. 6.2 Example of the filtering algorithm used on a minimum bias event contaminated with CW. The *left panel* shows the event averaged (*top*) and ϕ-sector averaged (*bottom*) f_P calculations, *red stars* indicate filtered bands and *dashed lines* indicate filter thresholds. The *right panel* shows the power spectrum in an antenna facing the CW direction before (*top*) and after (*middle*) filtering, then with *white noise* inserted in the filtered bands (*bottom*) (Colour figure online)

6.3.2 Interferometric Imaging

The key tool used in the ANITA-2 analysis, particularly for thermal noise rejection, is the interferometric image. Using data from all active channels in a given polarisation, the interferometric image provides information both on signal strength and the direction from which a signal originates as a function of payload coordinates.

By taking the cross correlation of waveforms from antenna pairs, it is possible to map out how well matched two waveforms are for a given time shift. The correlation coefficient returned here will be proportional to the amplitude of the two waveforms being processed, using normalised waveforms will provide a normalised correlation coefficient (Eq. 6.3.4).

$$C_{1,2} = \frac{\psi_1 \star \psi_2}{\sigma_{\psi,1}\sigma_{\psi,2}} \tag{6.3.4}$$

Here, $C_{1,2}$ is the normalised cross correlation between waveforms ψ_1 and ψ_2, while the normalisation of ψ_i is given by its RMS, $\sigma_{\psi,i}$.

If we treat RF signals reaching the ANITA-2 payload as plane waves then, for a given direction of incidence, there will be an offset between the time of arrival of the signal between any two given antennas. This timing difference varies as a function angle of incidence on the payload; as such the analysis calculates the correlation coefficient between two antennas as a function of azimuth (ϕ) and elevation (θ).

A global interferometric map can be constructed over all angles by summing and normalising correlation coefficients from all antenna pairs, as shown in Eq. 6.3.5.

$$C(\theta, \phi) = \frac{1}{n} \sum_{i \neq j} \frac{\psi_i \star \psi_j[\tau_{ij}(\theta, \phi)]}{\sigma_{\psi,i}\sigma_{\psi,j}} \tag{6.3.5}$$

Here, n is the number of channels used for a given direction, while τ_{ij} is the difference in arrival time of a plane wave between channels i and j incident at payload angle θ and ϕ. Each antenna will only contribute to a map if it is within 45° of the azimuthal pointing direction, and antenna pairs are only correlated if they are ≤ 2 ϕ-sectors apart. This allows for five ϕ-sectors worth of antennas to contribute to any one point on the map, while retaining only those antennas with a reasonable response to a given direction.

For the main analysis, the global payload coordinate interferometric image is constructed using 2° binning, as shown in Fig. 6.3. A refined 12° × 6° image, using 0.1° binning, is constructed for the direction of the peak of the coarse map for each event. Interpolation is used to extract a peak correlation coefficient, P_1, as well as relevant azimuth and elevation pointing angles, ϕ_1 and θ_1 respectively. From the global (coarse-binned) map, the correlation coefficient of the second largest peak, P_2, is taken from outside of a 10° × 10° exclusion region around P_1.

6.3.3 *Coherently-Summed Waveform*

If the direction of an incident RF signal is known, then a coherently-summed (that is, summing in phase) waveform from multiple channels will provide significant increase in signal strength compared to the waveform from a single channel. Meanwhile, summed waveforms for thermal events, for which there is no incident direction, will not sum coherently. This method, also known as beamforming, is used to provide a second discriminator between thermal noise and RF signals.

Using the event pointing angles from the interferometric image, θ_1 and ϕ_1, the coherently-summed waveform is created for every event. Appropriate timing shifts are applied to each channel within two ϕ-sectors of the event direction, with the shifted waveforms then added together and normalised by the number of channels used. Figure 6.4 shows coherent waveforms and their envelopes for an example calibration signal and a minimum bias event. A coherently-summed waveform is also constructed for each event using waveforms that have had the instrument response

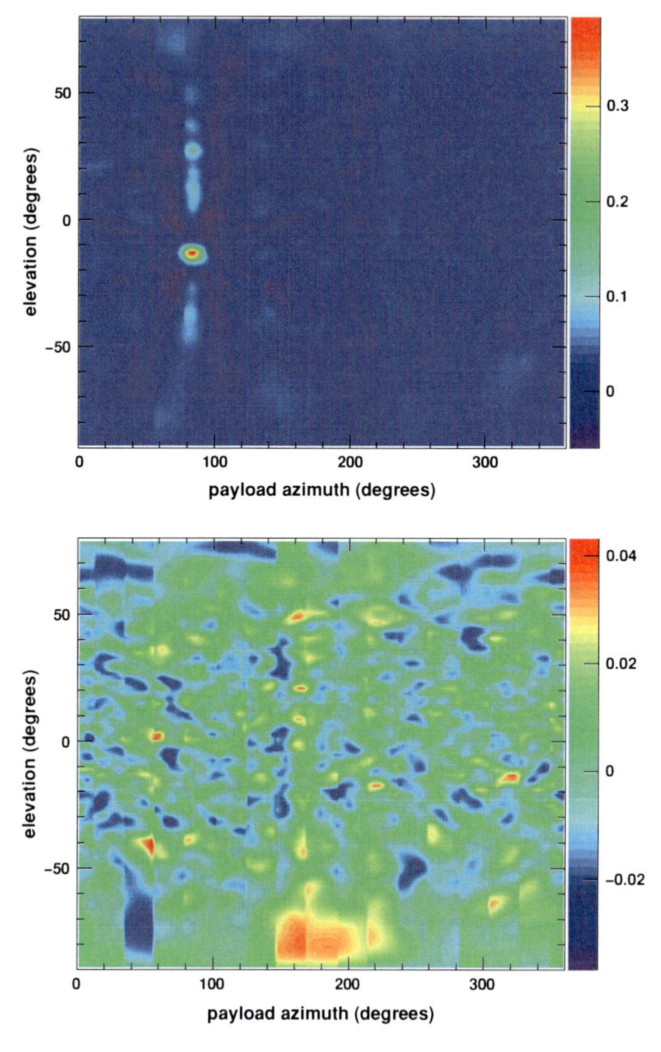

Fig. 6.3 Interferometric images for an event from the Taylor Dome calibration antenna (*top*) and a minimum bias trigger (*bottom*). Directionally coherent events, such as signals from the Taylor Dome calibration antenna, will display a unique peak in the interferometric image corresponding to the signals direction of incidence on the payload. Thermal events do not display this feature. The z-axes are on normalised scales; stronger or more coherent events will display higher peak correlation coefficients, with a value of one indicating perfect correlation between all channels

(see Figs. 4.7, 4.14) removed via deconvolution. This waveform is not used in the main thermal cuts.

The Hilbert envelope ($\Psi(t)$) of the coherently-summed waveform ($\psi(t)$) is found using Eq. 6.3.6.

$$\Psi(t) = \sqrt{\psi^2(t) + H^2(t)} \tag{6.3.6}$$

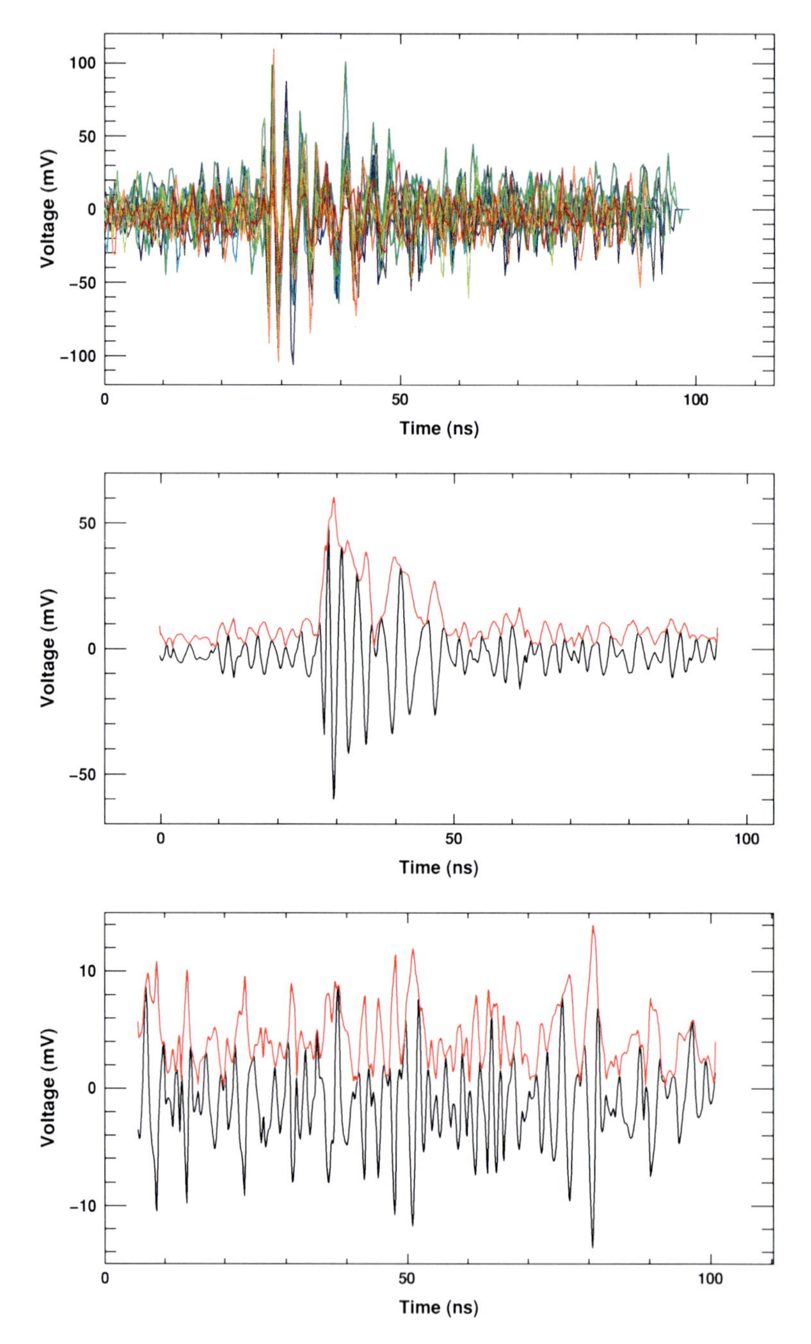

Fig. 6.4 Aligned VPOL waveforms for the channels in the five ϕ-sectors closest to the direction of incidence (*top*) and the resultant coherently-summed waveforms for an event from the Taylor Dome calibration antenna (*middle*). The coherently-summed waveform for a minimum bias trigger is also shown (*bottom*). The coherently-summed waveforms are shown with a *black line*, the Hilbert envelopes of the waveforms are shown with a *red line*

where the Hilbert transform $H(t)$ is the Fourier transform of $h(f)$, which is given by Eqs. 6.3.7 and 6.3.8.

$$h(f)(real) = -\psi(f)(\text{im}) \tag{6.3.7}$$

$$h(f)(im) = \psi(f)(\text{real}) \tag{6.3.8}$$

Equivalently, given the analytic signal, a complex function whose real component is $\psi(t)$, the Hilbert envelope $\Psi(t)$, is the amplitude of the analytic signal and so traces the envelope of this signal. For the analysis, two parameters are taken from the coherently-summed waveform's Hilbert envelope; these are the peak value of the envelope (H_P) and the time of this peak (H_t). Additionally, the coherently-summed waveform can be useful in detecting low level CW that is not apparent in single channel waveforms.

6.4 Thermal Cuts

The analysis described here aims to reject thermal noise via a small number of cuts, all of which are taken from either the interferometric image or the Hilbert envelope of the coherently-summed waveform. Because the expected number of signal events is very low, with even the most optimistic fluxes (ignoring exotic models) providing $O(1)$ neutrino event per flight, it is reasoned that the analysis should aim to remove all thermal background with confidence. As such, when setting thermal cuts the analysis will have a target thermal background rate of half of one event from the entire data set passing thermal cuts for each polarisation.

The three thermal noise samples are each useful in training thermal rejection cuts. Real thermal noise events, from the minimum bias trigger and upward-pointing event sample, would require extrapolation of cut values in order to provide the thermal noise rejection level required. As such, the final thermal rejection cuts in this analysis are trained on simulated noise events when possible, after it has been shown that the distributions of simulated events follow those of real noise events in the chosen cut parameters.

The Coherent Sun

It was noticed during the process of choosing thermal cuts that the P_1 values taken from the ANITA-2 noise samples were not in agreement with the those taken from simulated noise. Even after filtering and removal of self-triggered blast type events, the discrepancy remained. Figure 6.5 demonstrates that P_1 in the real data was dependent on angular separation between the position of the Sun and the event pointing. The same effect was observed in ANITA-1 data analysis [3]. From this we can conclude

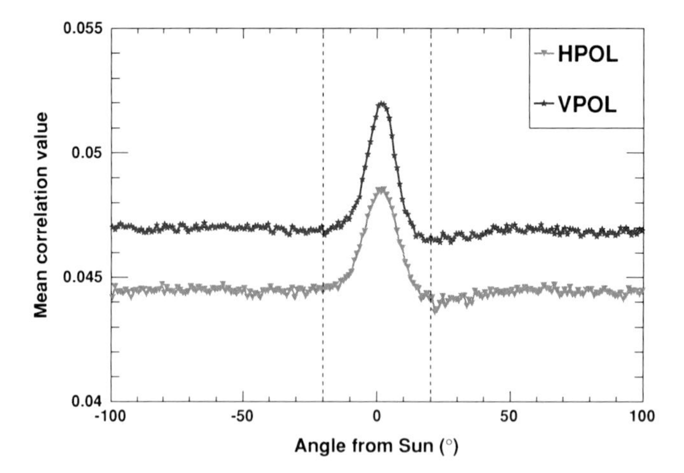

Fig. 6.5 Average peak correlation coefficient for minimum bias events as a function of azimuthal angular separation between event pointing and Solar position. *Dashed lines* indicate the $|\Delta\phi_S|<20°$ definition of whether an event points towards or not towards the Sun

that ANITA observes, and can resolve, coherent radio emission from the Sun. Figure 6.6 shows summed and averaged interferometric images with the Sun clearly resolved, with a reflection from the ice also visible in the HPOL image.

This effect prompted the division of the ANITA-2 analysis dataset into two separate samples for each polarisation before developing the final thermal noise cut, events pointing towards the Sun and events pointing away from the Sun. A conservative approach is taken when deciding whether an event points towards the Sun or not to remove the chance of any solar event leaking into the post cuts sample. Using Fig. 6.5, it was decided that any event with $\Delta\phi_S = |\phi_1 - \phi_{Sun}| < 20°$ would be defined as pointing towards the Sun.

As the simulated noise does not include the coherent Sun, the final thermal cut for the $\Delta\phi_S < 20°$ data is developed using real data. The final thermal cut for the $\Delta\phi_S > 20°$ data is trained using simulated data. The same thermal noise rejection level is used for both samples.

Peak Correlation

The interferometric image is constructed with a normalised correlation coefficient scale, meaning the P_1 and P_2 values of any two events can be directly compared. For directional events, P_1 should have a large significance over the correlation coefficients elsewhere in the interferometric image. For thermal events, the significance of P_1 is expected to be much lower. Figure 6.3 demonstrates the difference in P_1 between an incoherent thermal noise event and a coherent calibration signal.

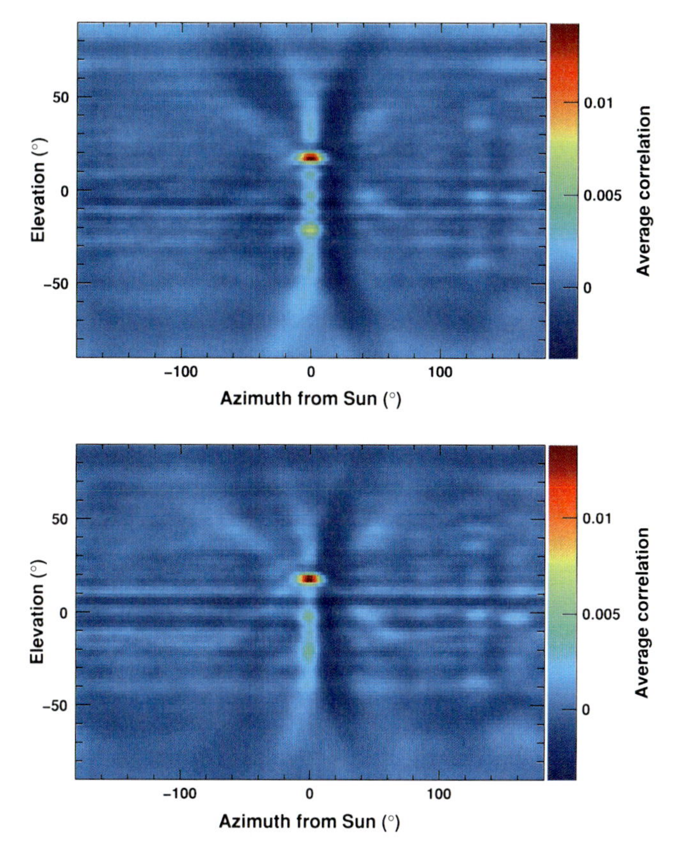

Fig. 6.6 Summed and averaged interferometric images for HPOL (*top*) and VPOL (*bottom*). Each image is constructed from 10^4 events, with the Sun clearly resolved in each

Simulated noise events display a peak correlation coefficient that is on average slightly lower than that of the minimum-bias and upward-pointing noise events with $\Delta\phi_S > 20°$. The peak correlation values from simulated noise are scaled by 1.025 in order to best match the real ANITA-2 data. The scaling value for this was found using a χ^2 minimisation, with the same scaling value used for both VPOL and HPOL events (Fig. 6.7).

A cut of $P_1 > 0.070$ is used. This removes a large fraction of thermal noise events, but would have been set lower were it not for the leakage of self-triggered blast events past the event quality cuts, described in Sect. 6.2. The cut was chosen such that no self-triggered blast in the upward-pointing noise sample passed all thermal cuts.

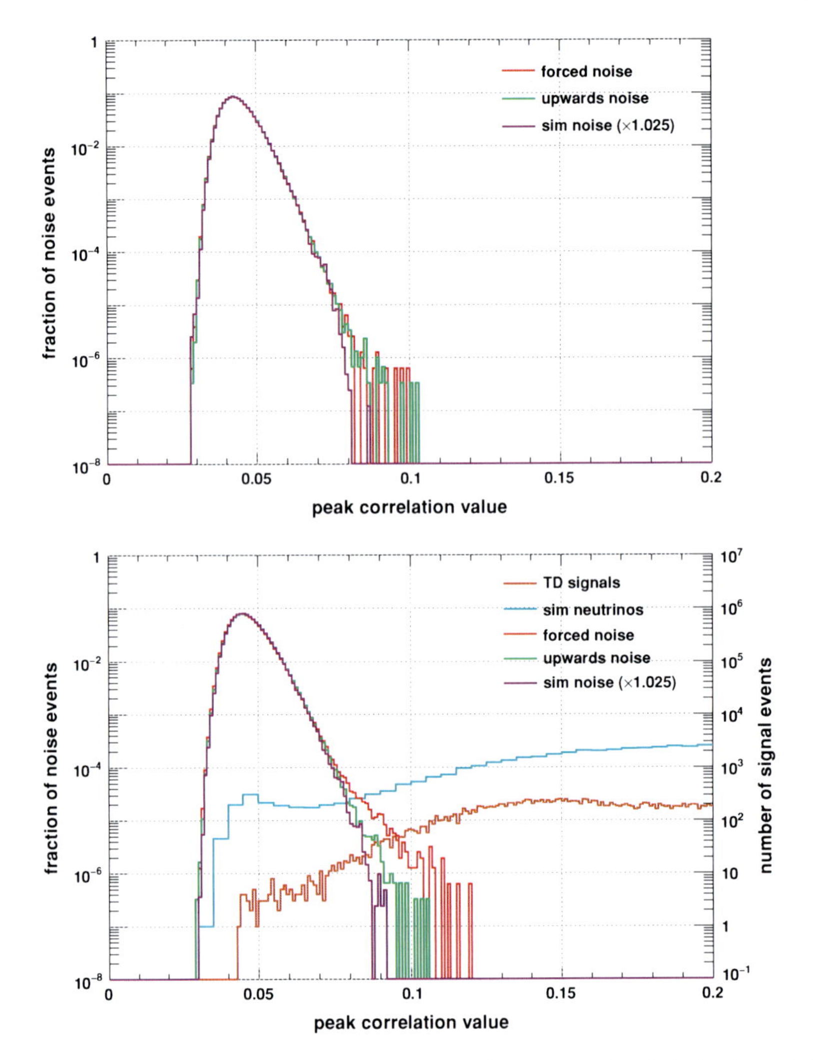

Fig. 6.7 Peak correlation coefficient for real (minimum bias and upward-pointing) noise and simulated noise events for HPOL (*top*) and VPOL (*bottom*). Data for VPOL signal-like events are also displayed for Taylor Dome calibration signals and simulated neutrinos. The peak correlation coefficients of simulated noise multiplied but 1.025 for the best match to real data, with tail distributions in VPOL caused by unfiltered CW contamination

Ratio of Correlation Peaks

A coherent event should display a clear and unique peak in an interferometric image, indicating that event's direction of incidence. While the absolute peak of the image, P_1, provides us with a measure of how coherent the event is in the given direction, the ratio of second to first peak correlation coefficient, P_2/P_1, allows for a test of

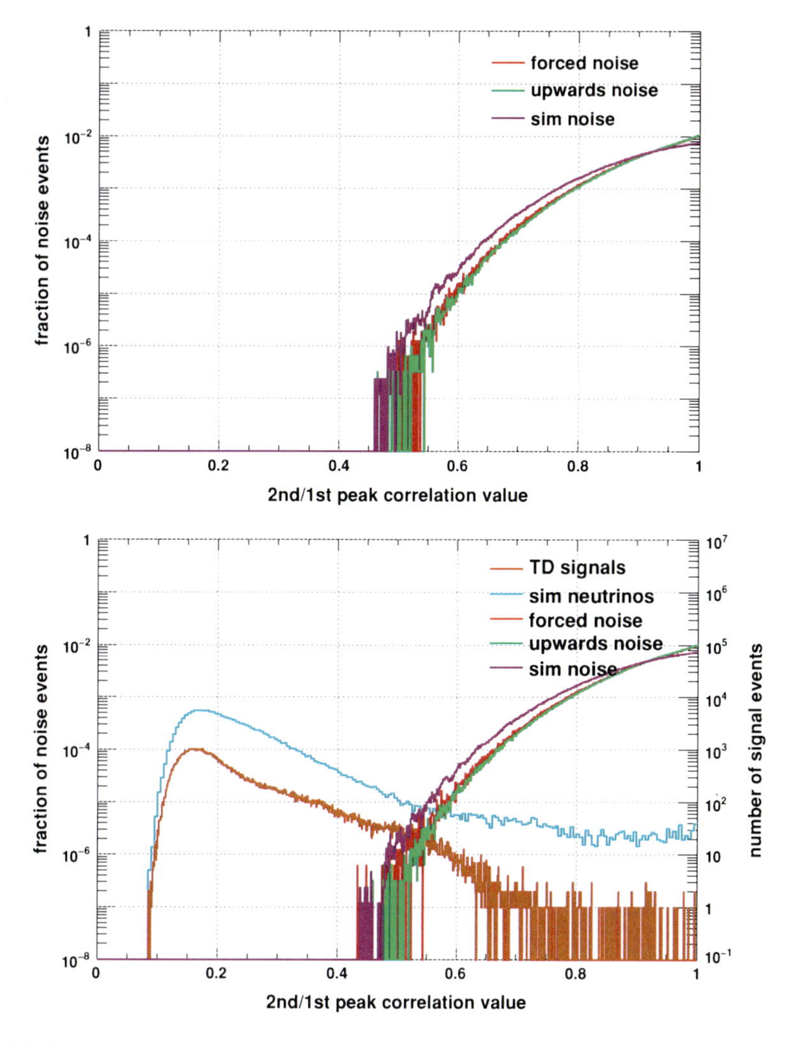

Fig. 6.8 2nd/1st peak correlation coefficients from the interferometric image for real (minimum bias and upward-pointing) noise and simulated noise events for HPOL (*top*) and VPOL (*bottom*). Data for VPOL signal-like events are also displayed for Taylor Dome calibration signals and simulated neutrinos

how uniquely the event points. P_2/P_1 provides good discriminating power against thermal noise events, as shown in Fig. 6.8, and can also be used as a test of whether an event has been misreconstructed, as these events will typically have a high P_2/P_1 value.

The P_2/P_1 cut is chosen on the basis that it retains all Taylor Dome events that are not misreconstructions and pass $P_1 > 0.07$. The chosen cut of $P_2/P_1 < 0.8$ satisfies both these criteria.

Event Elevation

The Seavey antennas used in ANITA-2 were highly directional, with a beamwidth of ~60°. The antennas were positioned with a downward cant of 10°, resulting in elevations of $< -40°$ being outside of all antennas' 3 dB point. The analysis accounts for off-boresight effects in antenna responses, however, any event reconstructing with $\theta_1 < -35°$ is cut from the analysis. Simulations show that this removes less than 1 % of ANITA-2's effective volume.

The upward noise events come from a sample with $\theta_1 > 0°$, these events are automatically excluded from analysis as they have been used to train analysis cuts. Therefore, all events reconstructing with $-35° \leq \theta_1 \leq 0°$ may pass thermal cuts, even though, with the horizon typically at $\theta_1 \sim 6°$, some portion of these events will not reconstruct to ground. See Sect. 6.5.1 for discussion of events that reconstruct above horizon.

Trigger Timing

The L3 trigger window is only ~10 ns in duration, while waveforms (and hence the coherent waveform) are ~100 ns long. The architecture of ANITA-2 means that when the TURF issues a trigger, the signal which resulted in the trigger command will be in the central region of the waveform data stored. A requirement is made that the section of the coherent waveform providing H_P is in rough agreement with the L3 trigger window. This ensures that the analysis is selecting signals that passed the hardware trigger.

To achieve this a cut is applied on the time of H_P, H_t, chosen such that all Taylor Dome impulsive events passing $P_1 > 0.07$ and $P_2/P_2 < 0.8$ also pass the H_t cut. The value chosen is 15 ns $< H_t <$ 70 ns, with a signal and noise comparison of the cut shown in Fig. 6.9.

Trigger Direction

To have agreement between analysis and hardware results, a cut is made on any events that reconstruct in a direction that does not agree with either a hardware trigger or software issued ϕ-mask. Any event with ϕ_1 in a direction that is not in, or adjacent to, a triggered or masked ϕ-sector is cut from the analysis. This corresponds to a ~22.5° cut on discrepancy between hardware and analysis directional information.

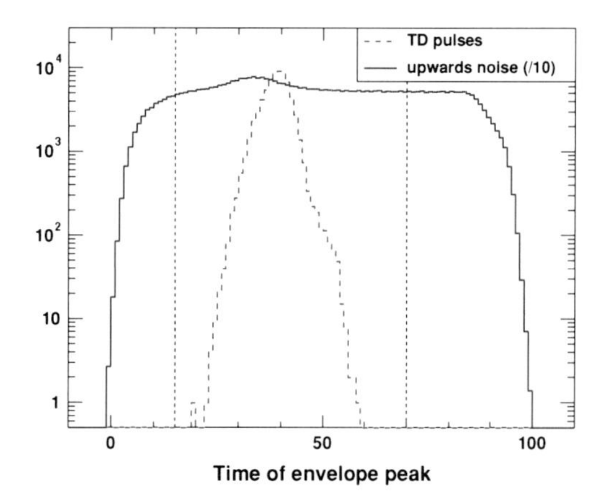

Fig. 6.9 t_{env} shown for upward-pointing RF-triggers and Taylor Dome Events (that do not misreconstruct). The bin contents of the RF-trigger histogram have been divided by 10

A number of RF triggered events display no L3 trigger,[1] as such, the analysis allows the trigger direction cut to use L2 triggers.

As mentioned previously, a decision was made to set thermal cuts such that 0.5 thermally-induced events would pass the analysis in the full data sample (2.1×10^7 events). This would provide us with confidence that events passing to the next stage of analysis contain coherent signals while also allowing for a sensitive analysis that would still select the majority of weak directional events.

To achieve this goal, it is necessary to apply all thermal cuts in sequence to a test (thermal-like) sample, such that an estimation of our expected thermal background can be made. As simulated thermal events are used for the training of final analysis cuts, the effect of each cut must be tested on, and a comparison made between, the simulated and real data samples. However, it is impractical to simulate enough thermal data to be able to test a trigger coincidence condition. Not including this cut in our final background estimate will both introduce additional uncertainty in the value obtained and result in more stringent values for other cuts, both of which are undesirable.

To remedy this, the trigger coincidence cut is tested on real data. Figure 6.10 demonstrates that, for incoherent events, the trigger coincidence cut removes 75 % of upwards-pointing thermal events, regardless of their peak correlation coefficient value. Only the non-thermal tail (primarily of satellite origin) of the upward-pointing triggers does not follow this trend. Therefore, a rejection value of 75 % is assumed for the trigger coincidence cut when training thermal cut values with the simulated (untriggered) noise.

[1] RF triggered events that display no L3 trigger were, in fact, L3 triggered events, but a rare condition in the FPGA trigger latching resulted in the L3 trigger information being recorded as negative in all ϕ-sectors.

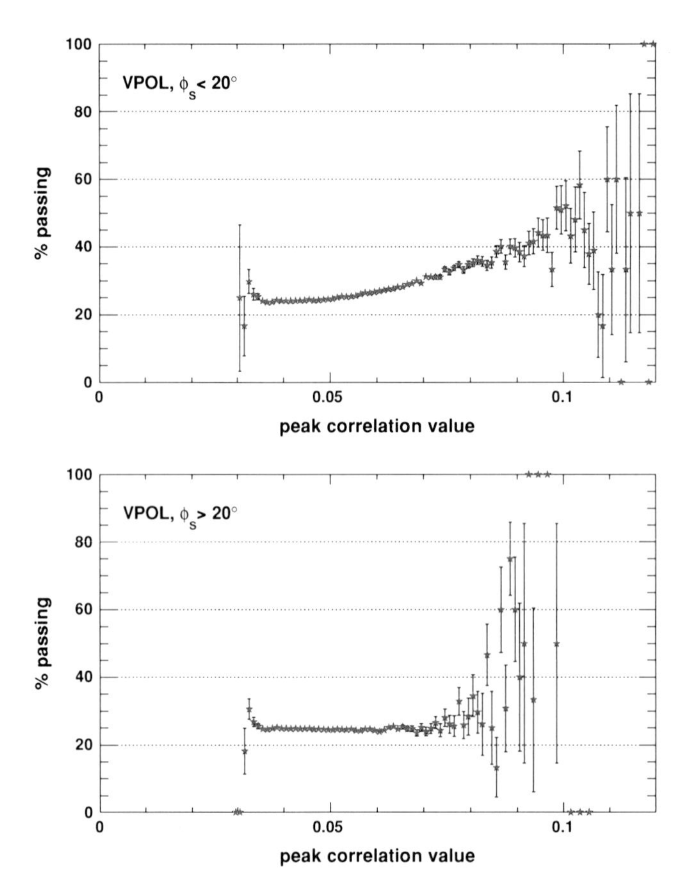

Fig. 6.10 Fraction of events passing the trigger coincidence cut as a function of peak correlation coefficient, *top* for VPOL events pointing $\Delta\phi_S < 20°$, *bottom* for events pointing $\Delta\phi_S > 20°$

Combination Cut

The final cut applied to remove thermal noise events is a combination of H_P and P_1 in the form $H_P + a_p P_1$. This was a method first developed by Stephen Hoover for an analysis of the ANITA-1 data [3] and provides significantly more discriminatory power between thermal and directional events than two separate cuts. Comparisons between signal and noise, and between simulated and real noise, for P_1 and H_P are given in Figs. 6.7 and 6.11, respectively. A comparison between H_P and P_1 is given in Fig. 6.12.

As the analysis data set has been divided according to where the event points in relation to the Sun, a different combination cut is chosen for each sample. For each data sample each of the previous cuts is applied initially. Simulated noise and Taylor Dome signals were tested with various values of a_p to find the value that maximised efficiency on Taylor Dome events for a given level of noise rejection.

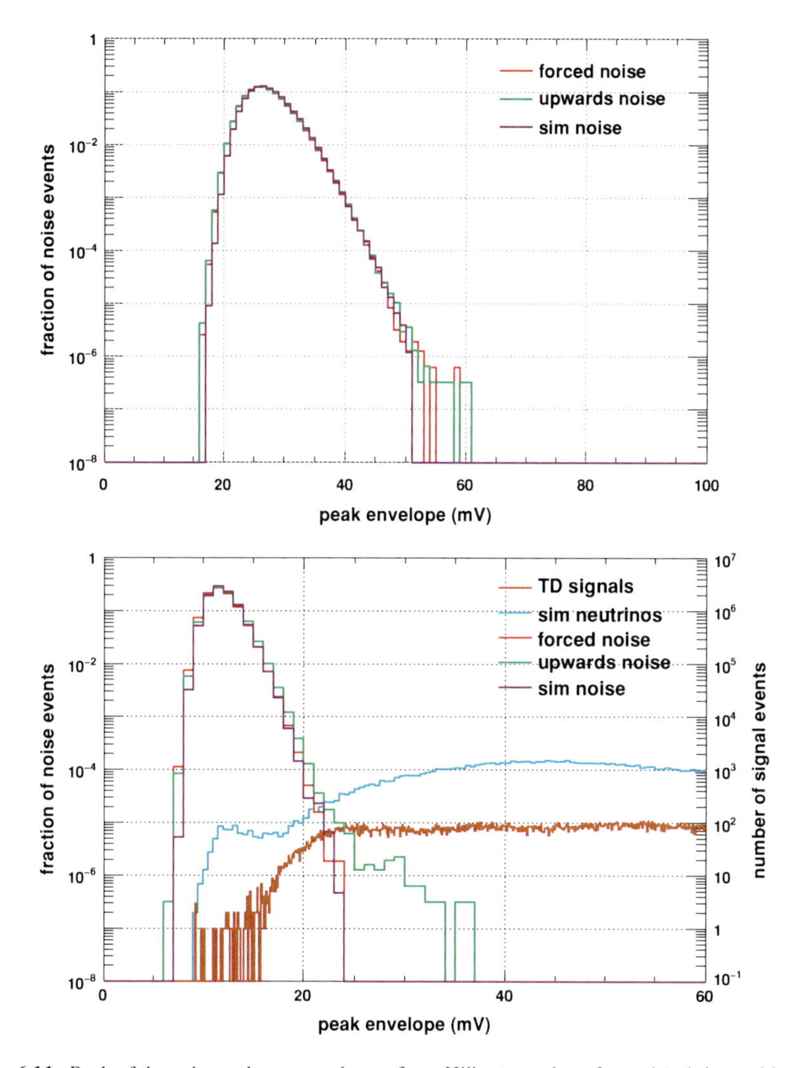

Fig. 6.11 Peak of the coherently-summed waveform Hilbert envelope for real (minimum bias and upward-pointing) noise and simulated noise events for HPOL (*top*) and VPOL (*bottom*). Data for VPOL signal-like events are also displayed for Taylor Dome calibration signals and simulated neutrinos. Excess in VPOL upward-pointing events over the other forced and simulated noise is caused by non-thermal events

From these tests, the value of a_P was set to 270. The final combination cut was chosen by requiring a noise rejection that would allow 2.5 events in 10^8 to pass all thermal cuts, corresponding to 0.5 events for the $\sim 21 \times 10^6$ RF events recorded by ANITA-2.

For events with $\Delta\phi_S \geq 20°$, simulated thermal noise is used to train the combination cut, for events with $\Delta\phi_S < 20°$, the upward-pointing noise sample is used. As

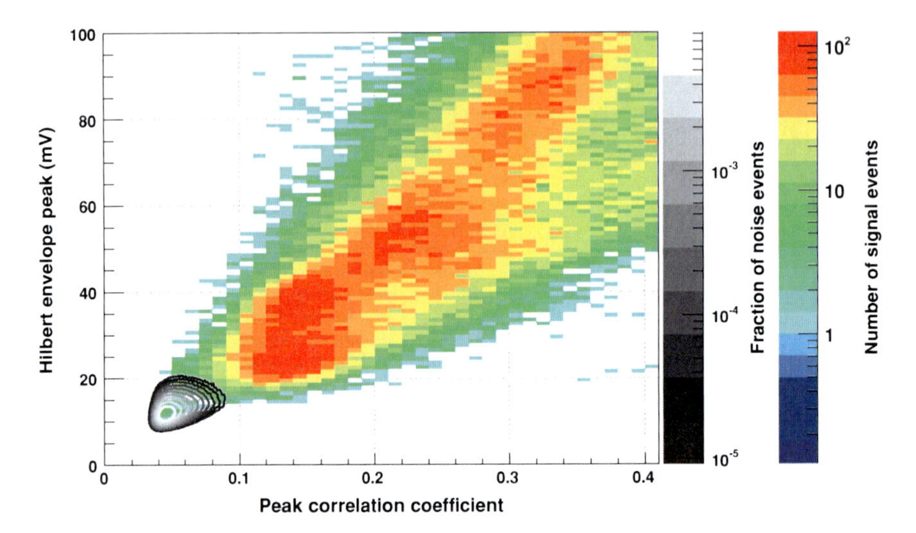

Fig. 6.12 Comparison of Hilbert envelope peak (H_P) and peak correlation coefficient (P_1) for VPOL events from Taylor Dome calibration signals (colour histogram) and upward-pointing thermal noise (*grey* contours). Note that no requirement is placed on pointing angle from the Sun for the either set of events in this figure

the upward-pointing noise events only represent a fraction of the ANITA-2 analysis data sample, the final cut is extrapolated by assuming a power law fit to the fraction of thermal events passing this final cut value.

Final cut values of $H_p + 270P_1$ for the VPOL analysis are 41.40 for events with $\Delta\phi_S \geq 20°$ and 61.36 for events with $\Delta\phi_S < 20°$. Final cut values of $H_P + 270P_1$ for the HPOL analysis are 64.81 for events with $\Delta\phi_S \geq 20°$ and 81.17 for events with $\Delta\phi_S < 20°$.

An error on the expected background of thermal events is calculated using the error in the fits to the fraction of simulated and thermal noise passing the combination cut from Figs. 6.13 and 6.14. The fits shown are of the form $A \cdot e^{b(x-x_0)}$. Fixing all parameters other than b, the expected thermal background passing thermal cuts is $0.50^{+0.27}_{-0.18}$ HPOL and $0.50^{+0.29}_{-0.18}$ VPOL.

6.4.1 Thermal Cut Results

A summary of the results of the thermal cuts on the analysis dataset is given in Table 6.3 for HPOL and Table 6.4 for VPOL events. The sample passing thermal cuts contains 267,272 events, with 66.67 % of the events being VPOL dominated.

A number of events passed thermal cuts in both VPOL and HPOL. Events which passed the thermal analysis in both polarisations but with the direction of reconstruction differing by $>3\sigma$ of the pointing resolution (see Sect. 6.5.1) between the two

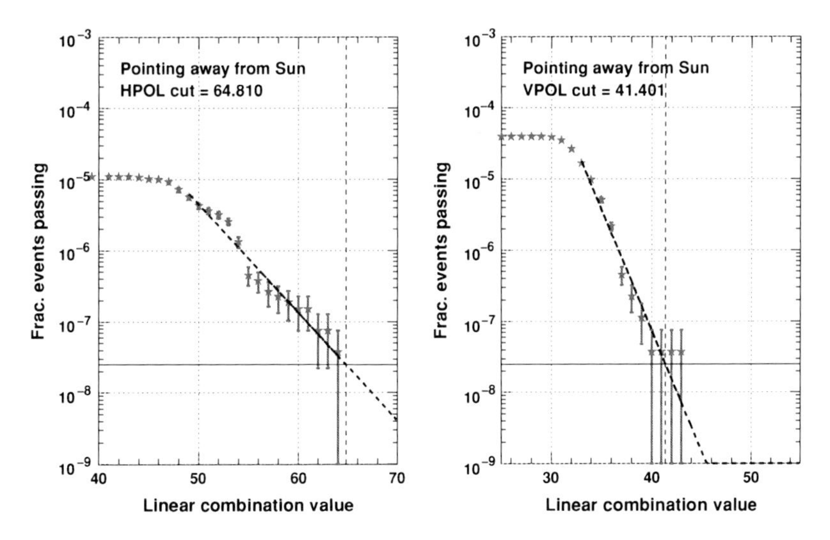

Fig. 6.13 The linear combination cut for HPOL (*left*) and VPOL (*right*) for simulated noise events pointing >20° from the Sun

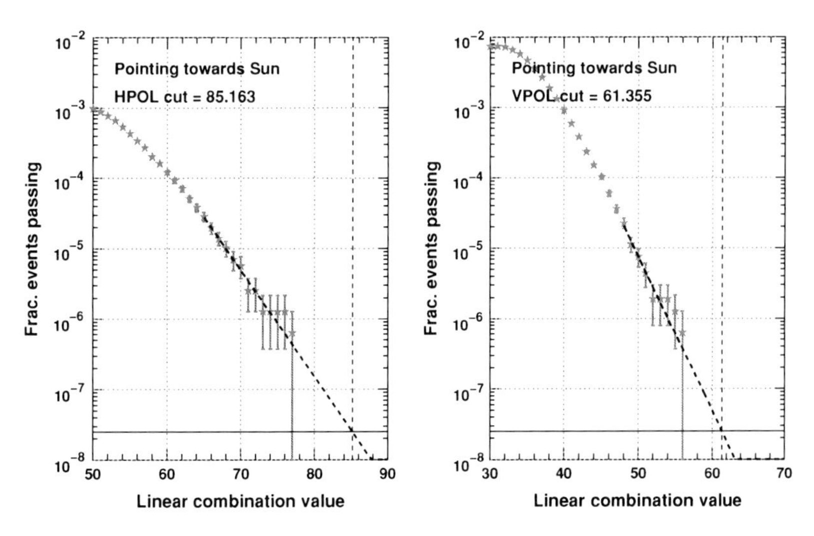

Fig. 6.14 The linear combination cut for HPOL (*left*) and VPOL (*right*) for upward-pointing noise events pointing <20° from the Sun

polarisations were cut from the analysis (labelled as "point differently" in Tables 6.3 and 6.4). Only one polarisation was used for all other events which passed the thermal analysis in both polarisations, with the polarisation selected as that which had the highest P_1 coefficient (labelled as "point better" in Tables 6.3 and 6.4). Finally, any event passing either polarisation may not reconstruct below 0° in elevation but above

Table 6.3 HPOL cuts summary

Cut	Fails	In seq	If last	Pass without
$P_2/P_1 < 0.8$	19275570	19272586	2818	213303
$P_1 > 0.07$	20837426	1604424	773	211258
$270P_1 + H_P$[a]	20930699	68653	19959	230444
$\theta < 0°$	10390446	42	15	210500
$\theta > -35°$	3354611	56724	55647	266132
$15\,\text{ns} < H_t < 70\,\text{ns}$	7473923	6395	6366	216851
$\lvert\phi$-trigger/mask$\rvert < 2$	14313518	278	278	210763
			Cut	Remaining
				210485
Point differently V			15	210470
Point better V			121431	89039
Doesn't reconstruct to ground			2	89037
Events to cluster				89037

[a]The combination cut $270P_1 + H_P$ depends on the angle to the Sun

Table 6.4 VPOL cuts summary

Cut	Fails	In seq	If last	Pass without
$P_2/P_1 < 0.8$	19021767	19019369	1448	261890
$P_1 > 0.07$	20626758	1701693	345	260787
$270P_1 + H_P$[a]	20789328	145236	20089	280531
$\theta < 0°$	10530819	1184	106	260548
$\theta > -35°$	3086198	61927	60181	320623
$15\,\text{ns} < H_t < 70\,\text{ns}$	7101849	29074	28787	289229
$\lvert\phi$-trigger/mask$\rvert < 2$	14315471	662	662	261104
			Cut	Remaining
				260442
Point differently H			15	260427
Point better H			82098	178329
Doesn't reconstruct to ground			94	178235
Events to cluster				178235

[a]The combination cut $270P_1 + H_P$ depends on the angle to the Sun

the horizon. Section 6.5.1 describes which events from this sample are labelled as "doesn't reconstruct to ground" in Tables 6.3 and 6.4 and cut from the analysis.

6.5 Anthropogenic Cuts

The majority of the 267,272 events that passed thermal analysis cuts will have been caused by anthropogenic noise. Such events will usually be associated spatially with either a known site of human activity, or with other events of a similar origin. To reject these events, a clustering algorithm is developed that considers all events passing thermal cuts and groups associated signals together. For clustering to be reliable the algorithm will require knowledge of ANITA-2's pointing resolution along with a comprehensive set of data on human activity in Antarctica during ANITA-2's flight.

6.5.1 *Pointing Resolution*

The pointing resolution of the analysis was tested using Taylor Dome calibration signals. As the Taylor Dome calibration antenna was vertically polarised, the pointing resolution is only assessed for VPOL event analysis. Although calibration pulses from the Williams Field Seavey antenna were transmitted in HPOL, the event sample is small (<10000 events), is heavily contaminated with CW and does not cover a suitable range of signal strengths to allow us to assess ANITA-2's HPOL pointing resolution. We therefore assume that the HPOL pointing resolution is similar to that in VPOL.

Figure 6.15 shows the VPOL pointing resolution as a function of peak correlation coefficient, calculated using every Taylor Dome calibration signal that passed thermal analysis cuts. The resolution, σ, is calculated for an angle, i, using the root-mean-square (RMS):

$$\sigma_i = \frac{1}{n}\sum_{i=0}^{n}\left(\Delta(i^2) - (\Delta(i))^2\right) \qquad (6.5.1)$$

where $\Delta(x)$ is the difference between the true and measured value of x. Both azimuthal and elevation pointing resolutions are sub-degree for peak correlation coefficients of >0.1. Elevation resolution is roughly a factor of two better than azimuthal resolution due to the vertical antenna separation being larger than the horizontal one.

Using the pointing resolution, a directional uncertainty is assigned to every event passing thermal cuts for use in event clustering. There exist a small number of events that pass thermal cuts, but reconstruct above horizon. For these events we allow the elevation pointing angle to be moved downward until the event reconstructs to ground. If the change in elevation required for reconstruction is $\leq 0.5°$ or $\leq 2\sigma_\theta$ then the event is passed into the event clustering algorithm, otherwise the event is cut.

6.5.2 *Base and Flight Lists*

A list of human bases (both active and inactive) and automated weather stations was compiled for the ANITA-2 analysis [4]. This list is not assumed to be either complete or exhaustive. Further human bases are inferred from a radio coherence map, created by projecting the peak correlation coefficient of every event passing the quality cuts to ground and normalising over the number of events pointing to

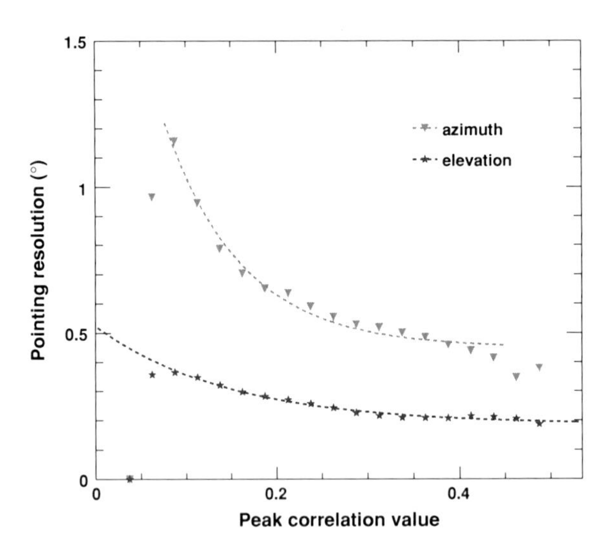

Fig. 6.15 Pointing resolution in azimuth (ϕ) and elevation (θ) as a function of peak correlation coefficient for VPOL signals from the Taylor Dome calibration antenna. Resolution is found via the RMS, given by Eq. 6.5.1

each location, results are shown in Fig. 6.16. For the construction of this map event filtering was turned off. The results can be compared with similar maps created with filtering implemented, shown in Fig. 6.17. Any location with an averaged correlation value of >0.070 was assumed to be associated with anthropogenic noise. A peak finding algorithm searched these areas and placed a 'pseudo-base' at the location of a local maximum correlation value, these pseudo-bases were included in event clustering and treated as normal bases.

During the 2008–2009 austral Summer, a number of scientific programs made use of aircraft with on-board radar transmitters and receivers. Information on these flights, with relevant GPS and timing information, was obtained for use in the ANITA-2 analysis [4]. Further to this, three traverses across the Antarctic continent were made for which position and timing information were also obtained. While static bases and pseudo-bases are used in anthropogenic noise rejection for the entire non-thermal data, flight GPS position information is only used if the GPS timing and ANITA-2 event timing coincide to within half an hour. For the traverses, the positional data is used for anthropogenic rejection if it was recorded within one day of a given ANITA-2 event.

6.5.3 Clustering Algorithm

Two clustering cuts are used in the analysis that complement each other. A simple distance cut groups events and bases together if they are separated by <40 km. This cut is useful at smaller balloon-to-source separations, while at large balloon-to-source separations an angular error will translate to a larger separation error. To account for

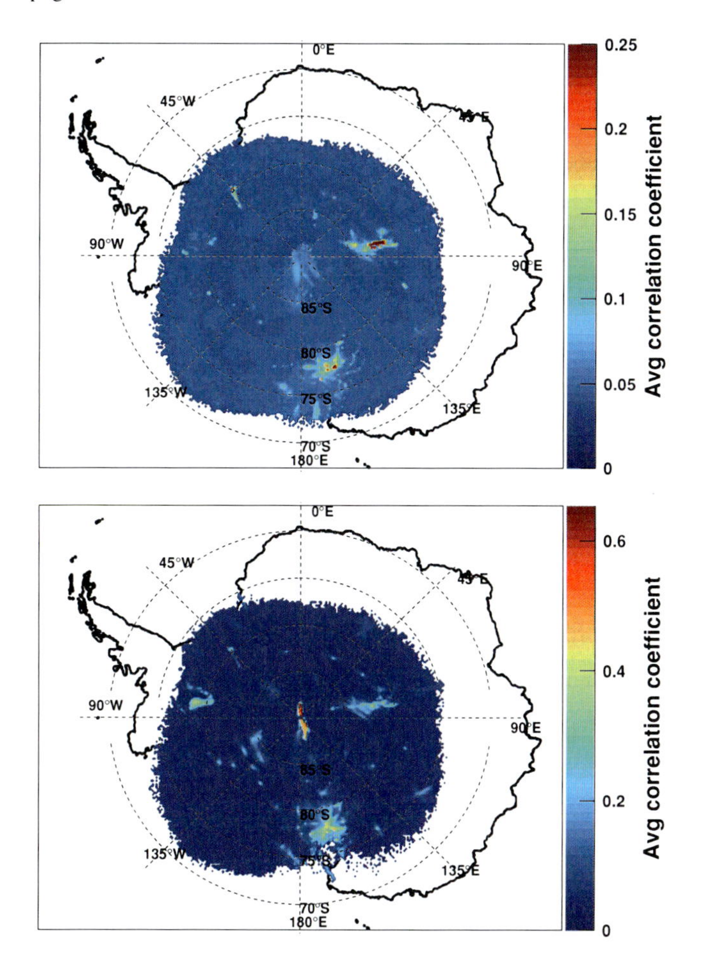

Fig. 6.16 Maps of the average correlation value of every event passing quality cuts for HPOL (*top*) and VPOL (*bottom*) with no event filtering

this, a log-likelihood metric is used to cluster events with one another using the event pointing resolution:

$$-2LL = \left(\frac{\theta_{Aa} - \theta_{Ab}}{\sigma_{\theta a}}\right)^2 + \left(\frac{\phi_{Aa} - \phi_{Ab}}{\sigma_{\phi a}}\right)^2 + \left(\frac{\theta_{Ba} - \theta_{Bb}}{\sigma_{\theta b}}\right)^2 + \left(\frac{\phi_{Ba} - \phi_{Bb}}{\sigma_{\phi b}}\right)^2 \tag{6.5.2}$$

Here the A and B subscripts indicate the balloon location, a and b indicate the location of the associated reconstruced events and $\sigma_{\alpha j}$ is the resolution in angle α for event j. So, θ_{Ii} is the measured elevation angle for event i, while θ_{Ij} is the projected elevation angle for event j onto the balloon's location and orientation when recording event i. For the case of event to base clustering, there is only one relevant balloon

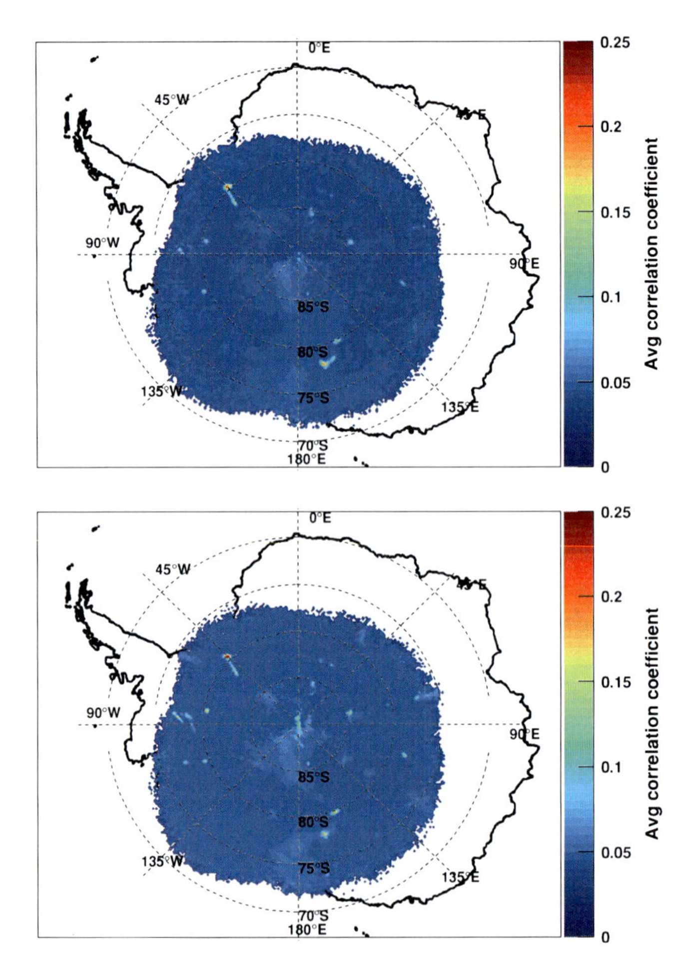

Fig. 6.17 Maps of the average correlation value of every event passing quality cuts for HPOL (*top*) and VPOL (*bottom*) with event filtering used

position and associated event, so the expression simplifies to:

$$-2LL = \left(\frac{\theta_{event} - \theta_{base}}{\sigma_\theta}\right)^2 + \left(\frac{\phi_{event} - \phi_{base}}{\sigma_\phi}\right)^2 \tag{6.5.3}$$

Using both the Williams Field and Taylor Dome borehole antennas, the fraction of events being associated with the relevant base location is plotted in Fig. 6.18. A cut on the likelihood value of $-2LL = 30$, corresponding to a pointing $\sigma \sim 5.5$, is used for event clustering.

An analysis by Abigail Vieregg of the ANITA-2 data, summarised in [5], returned a significantly larger number of events than this analysis after the thermal cut stage.

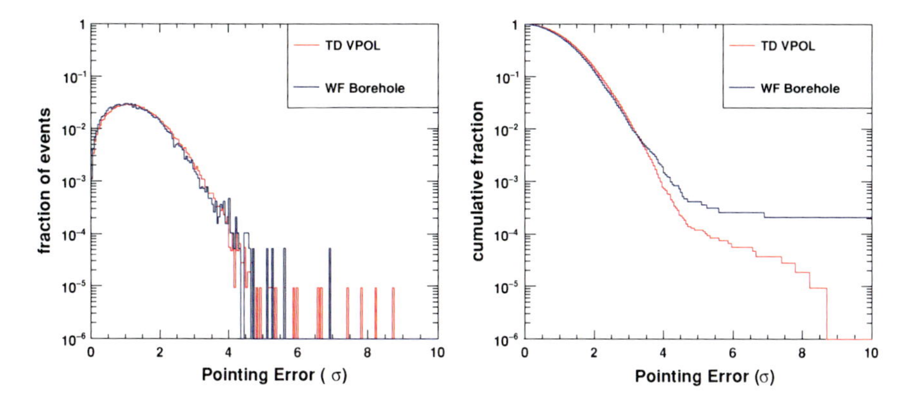

Fig. 6.18 *Left* The fraction of calibration signal events as a function of pointing error, $\sigma = \sqrt{(2LL)}$. *Right* The cumulative fraction of events being clustered as a function of pointing error. Note that at $\sigma > 5$ events are heavily contaminated with CW

On comparing events failing thermal cuts from this analysis that passed thermal cuts in [5], it was noticed that the filtering used in this analysis was considerably stronger, removing much weaker CW contamination. While this is useful when analysing impulsive events, as potential contamination is removed more successfully, it is also possible that there are CW sources not included in the base list or the coherence map analysis that [5] did include. It was decided that the locations of events from [5] that failed thermal cuts in this analysis should be included in the clustering stage and removed before unblinding. The net effect of this is to reduce our neutrino efficiency, but it also reduces the risk of impulsive events associated with weak CW sources being claimed as signal events.

The clustering algorithm first tests all events for association to flights and traverses. Any event with a separation <40 km or likelihood $-2LL < 30$ to a flight or traverse location is removed from the analysis, these events are not included in further clustering. Clustering then compares all events to every base as well as to every other event. It is possible for an event to be clustered with both, one of, or neither base or event. It is also possible for an event to be clustered to more than one base or other event. The algorithm finally sorts all the discovered associations, such that the analysis returns clusters of variable sizes that fall into one of two types: clusters associated with one or more bases (**base clusters**) and clusters not associated with a base (**event clusters**).

6.5.4 Clustering Results

A summary of the event clustering is shown in Table 6.5. All events passing thermal analysis cuts that cluster to either bases or other events are displayed in Fig. 6.19. By far the largest concentration of events originate from the region around McMurdo,

Table 6.5 A summary of the number of events clustered to flights/traverses and bases

Cluster type	Cluster size	From this analysis
Flight/traverse		527
Base clusters:		
McMurdo	348322	168619
Unknown pseudo base	175	5
AGO 1	66	3
AGAP South	161	161
Davis-Ward	28639	28406
Beardmore Camp	2	2
South Pole	117	117
Dome C	1765	54
Vostok	2941	2821
AGAP North	18	18
AGO 3	7785	897
Dome F	81	1
AWS Baldrick	2045	13
AGO 2	194	26
Patriot Hills	19327	18949
Mt. Takahe	1	1
AGO 5	155	155
Berkner Island	45028	44752
Belgrano II	1	1
Unknown pseudo base	1690	1649
Camp Neptune	6	6
Cordiner Peak	6	6

Clusters are named with the largest single partner base. Exact sizes of clusters of less than 100 events were not inspected prior to unblinding in order to hide the final analysis results. Note that the very large cluster assigned to McMurdo includes a number of clusters from nearby bases that were associated by the clustering algorithm

reflecting both the level of human activity in this region and the fact the ANITA-2 made three separate passes directly over the area.

6.6 Efficiency and Background

The efficiency of the analysis is calculated as a function of primary neutrino energy using simulated events. The results of event clustering are used to estimate the anthropogenic background. This is then combined with a post-clustering thermal background to provide the expected number of non-physics events that will appear as isolated signals after all analysis cuts.

Fig. 6.19 Map of all events passing thermal cuts but failing clustering cuts. Events marked in *green* are associated with one or more bases, events marked in *red* were in event only clusters, bases are indicated with *black markers*

6.6.1 Analysis Efficiency

The overall sensitivity of the analysis as a function of primary neutrino energy is shown in Fig. 6.20. The calculation of the overall efficiency includes three separate considerations: would the event pass event quality cuts, would the event pass thermal cuts and would the event pass clustering cuts.

It is found that simulated neutrinos passing the hardware trigger will only fail one of the quality cuts: digitiser (SURF) saturation. As would be expected, the proportion of events failing this cut increases with neutrino energy. As the non-linearity of the amplifiers has not been taken into account for strong signals, the values shown in Fig. 6.20 provide a conservative estimate on the fraction of simulated neutrino events that would result in digitiser saturation.

It has already been shown that the simulated hardware efficiency is slightly lower than the actual instrument sensitivity. As the thermal analysis efficiency is excellent for all but the weakest signals, the thermal analysis is very efficient for simulated neutrinos passing the hardware trigger. A decrease in thermal analysis efficiency with increasing energy is displayed in Fig. 6.20. Simulated neutrinos failing the thermal analysis contain a much larger fraction of their power at low frequencies, with the spectrum showing a steep peak in the 200–350 MHz region, as shown in Fig. 6.21. The filtering algorithm for these events is unable to distinguish the radiation from narrowband noise, the event is filtered and then fails thermal cuts.

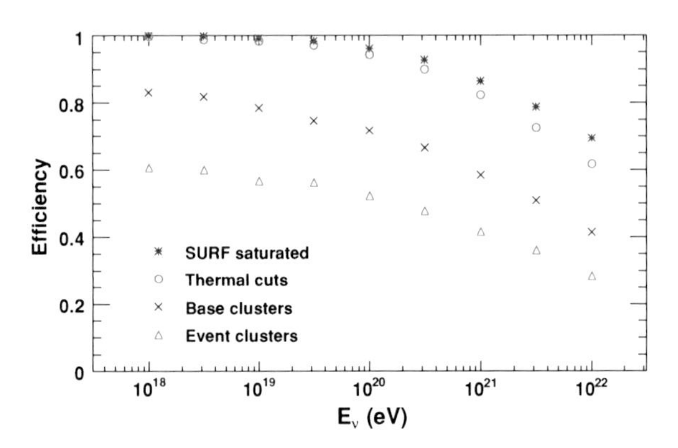

Fig. 6.20 Analysis efficiency to simulated neutrino signals

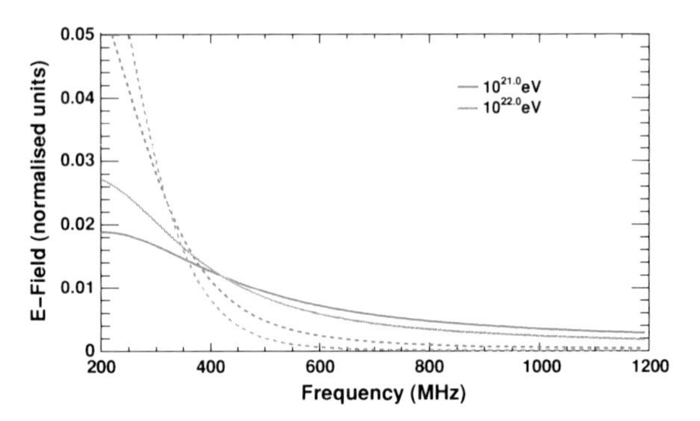

Fig. 6.21 The electric field at the payload for simulated neutrinos failing (*dashed*) and those passing (*solid*) thermal cuts at two different energies. Events are normalised and averaged, with >1000 events used for each spectrum. The events failing thermal cuts display a stronger peak at low frequencies, which resulted in much of their power being filtered

The clustering efficiency is the real limiting factor in analysis sensitivity to neutrino signals. At the time of the ANITA-2 flight a number of scientific programs were running at bases situated on some of the deepest ice in Antarctica. This resulted in a significant reduction in ANITA-2's detector volume through clustering cuts when compared with ANITA-1, though the experiment as a whole remained more sensitive to neutrinos.

6.6.2 Background Calculation

An expected background of thermal events has been set through the level of stringency chosen for the thermal cuts. Further background may arise from misreconstruction of coherent event locations. However, checks with the Taylor Dome calibration signals display a negligible misreconstruction rate for even weak impulsive events. Meanwhile, any misreconstructed CW event, while undesirable, should be easily identified and removed after unblinding. By far the most significant background for ANITA-2 data analysis is that of single anthropogenic events that do not originate from a known or pseudo base. While human activity in Antarctica is limited and usually well

Table 6.6 Anthropogenic background estimates

Events from [5]	Base clusters	Base singles	Event clusters	Expected singles
Treated as bases	10	3	3	0.90
	9.4 % HPOL, 90.6 % VPOL			0.08 H, 0.82 V
Removed from sample	13	5	4	1.54
	5.9 % HPOL, 94.1 % VPOL			0.09 H, 1.45 V

Events from [5] that fail the thermal cuts of this analysis are treated in two different manners in event clustering, providing two separate background estimates

documented, it is very possible that there exist small bases which are not included in the ANITA-2 base list. Additionally, not all aircraft activity over the continent is included in the flight lists used for event clustering.

To calculate the expected isolated event background from human activity, a key assumption is made that the distribution of signal strengths from small, known, bases is the same as the distribution of signal strengths from small, unknown, bases. Using this assumption, and defining small bases as those with <100 events originating from them, the expected anthropogenic background ($N_{single,non\text{-}base}$) can be estimated:

$$\frac{N_{single,non\text{-}base}}{N_{small\ cluster,non\text{-}base}} = \frac{N_{single,base}}{N_{small\ cluster,base}} \tag{6.6.1}$$

Using the fraction of events from each polarisation that make up the events providing the background calculation, we can then produce a separate background estimate for each polarisation. The number of small clusters, both from bases and non-base clusters, along with the number of base singles, are low. This leads to a relatively large uncertainty in the background calculation given in Eq. 6.6.1. Two treatments of the event sample are used, providing independent estimates of the anthropogenic background, with results shown in Table 6.6. The resultant expected anthropogenic background is 1.13 ± 0.32 VPOL and 0.09 ± 0.01 HPOL. The expected VPOL background is significantly higher due to all triggering being VPOL based, the vast majority of signals from small clusters are classed as VPOL by the analysis.

The clustering of simulated neutrinos, averaged over all energies, results in ~57 % of events being rejected as members of base or event clusters. The thermal background estimate of 0.50 events in each polarisation was made without considering this clustering efficiency. Applying the same level of efficiency to thermal background leaves an expected thermal background of $0.29^{+0.16}_{-0.11}$ HPOL and $0.29^{+0.17}_{-0.11}$ VPOL events. Combining this with the expected anthropogenic background, the overall background expectation for the analysis is $0.38^{+0.16}_{-0.12}$ HPOL events and $1.42^{+0.36}_{-0.34}$ VPOL events.

Table 6.7 Outcomes of analysis on the inserted Taylor Dome events

Event #	Thermal cuts	Anthropogenic cuts
4586631	Fail	–
8381355	Pass	Fail
9362397[a]	Pass	Pass
10345208	Fail	–
11207106	Pass	Pass
12943715	Pass	Pass
13662401[a]	Pass	Fail
15406954	Pass	Fail
19480638	Fail	–
20564174	Pass	Fail
24699044	Fail	–
25887362	Pass	Fail

[a]Events 9362397 and 13662401 are the same Taylor Dome event, inserted twice into the data

6.7 Events Passing Thermal and Anthropogenic Cuts

With an expected combined background calculation predicting of order 1 event in each polarisation (consistent with the initial aim of this analysis), the analysis signal region is unblinded. In this section, the analysis outcomes of the inserted Taylor Dome calibration signals are summarised and isolated singles passing thermal cuts are inspected.

6.7.1 Inserted Taylor Dome Events

A total of twelve Taylor Dome events were inserted into the ANITA-2 flight data. Only eleven of these events are unique, as one Taylor Dome event was inserted twice into the data. The outcome of the analysis in processing the inserted events is given in Table 6.7.

Of the eleven unique events to be inserted, seven passed thermal cuts. One of these was the duplicated event, so a total of eight events were passed in to the clustering algorithm. Of these eight events, three passed clustering cuts.

The measured efficiency on the inserted Taylor Dome events is consistent with both the thermal analysis efficiency from the cut training stage and with the clustering efficiency. Efficiency would have been improved slightly with a weakened combination cut on events pointing towards the Sun, or a tighter region from which events had the "towards Sun" cut applied. One of the inserted Taylor Dome events were assigned as pointing towards the Sun (event # 24699044). This event failed thermal cuts, but would have passed cuts had it pointed away from the Sun.

Table 6.8 A list of events passing all analysis cuts

Event #	Polarisation	Comment
3182655	HPOL	Strong CW
3478716	VPOL	Associated CW
14496361	HPOL	
14577785	HPOL	
14917312	VPOL	
15636066	VPOL	
16014510	VPOL	
21970905	VPOL	Payload Blast
22807894	VPOL	
23056816	VPOL	
26507672	VPOL	Payload Blast
27146983	HPOL	

6.7.2 Isolated Signals

After unblinding a total of twelve events were found to pass all thermal and anthropogenic cuts, these events are listed in Table 6.8.

A number of these events are of a type that should have been removed by event quality or CW misreconstruction cuts. Looking at each of the isolated events in turn, events are removed if it is possible to provide a convincing reason as to their unsuitability as a physics candidate event.

Table 6.8 shows that four events in total are removed by hand due to easily recognisable poor event quality. One of these is removed due to strong CW in all HPOL channels that resulted in event misreconstruction. One event is well reconstructed, however, there is a strong CW signal in the few seconds before and after the event triggered the same ϕ-sector and appears to reconstruct to the same source location. Two events have signatures of self-triggered blasts, but were not removed by event quality cuts.

6.8 Impulsive Signal Cut

It was noticed after unblinding that a number of events passing the thermal cuts, while reconstructable, had no impulsive nature to the coherent waveform. The approach of the analysis until this point was to place no constraints on overall signal shape, as predictions of Askaryan signals from UHE neutrino interaction remain uncertain. However, we do expect there to be some impulsive signal that should be resolvable to some extent in a waveform that does not contain the instrument response, regardless of features that, for example, the LPM effect could introduce. Using the signal chain and antenna response (see Figs. 4.7, 4.14), an instrument-response-deconvolved,

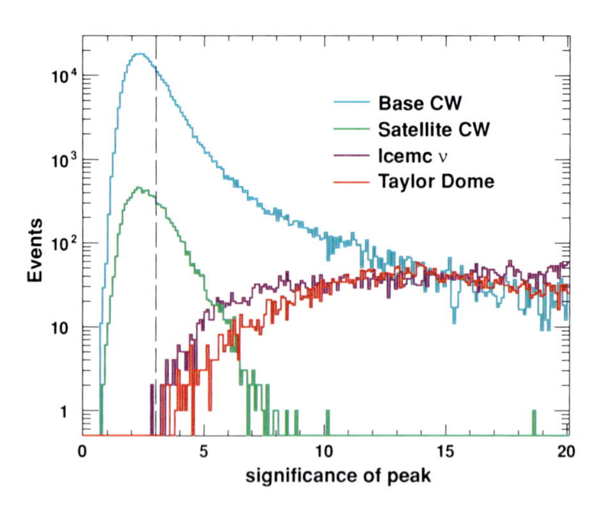

Fig. 6.22 The result of the impulsive signal requirement cut on satellite CW, base CW, Taylor Dome calibration signals and simulated neutrino signals. Taylor Dome and simulated neutrino events are assessed with regular analysis filtering and are only considered if they pass all previous thermal cuts. Satellite and base CW events are assessed with filtering turned off, only satellite CW events passing all previous thermal cuts are included in the plot. The *vertical line* denotes the cut value selected for use in analysis

coherently-summed, waveform can be constructed for each event using the same method described in Sect. 6.3.3.

It was decided to produce a final physics candidate sample by applying a test of impulsive signals using the instrument deconvolved coherent waveform. In order to remove as much bias as possible, the impulsivity test was trained using the Taylor Dome calibration signals and simulated neutrinos as signal samples and non-impulsive reconstructed events as the background sample. The background sample was created by removing filtering from the analysis and selecting upward-pointing events from three runs of data when ANITA-2 was far from radio-loud human bases. All upward-pointing events that pass all other thermal cuts were used and should originate from satellite noise. The sample of anthropogenic events that passed the analysis outlined in [5] but failed this analysis (largely due to filtering) were also tested, with an unfiltered, coherently-summed, waveform constructed using pointing coordinates from [5].

Impulsiveness was tested by passing a 5 ns window across the coherently-summed, instrument-response-deconvolved, waveform and averaging the power within the window. A peak value of the 5 ns averaged power was forced to come from between 23 and 63 ns of the trigger time. Average power values were calculated for the periods prior to the earlier and after the latter timing cut. The significance of the peak was then taken as the larger of peak/average using either pre- or post- window averages. The timing cuts were selected by running an optimisation of the Taylor Dome and satellite signals that passed all thermal cuts.

It was found that a cut of peak/average > 3 retained all Taylor Dome signals passing thermal cuts, while removing 66 % of both background samples. One simulated neutrino, from a total sample of $>$40,000, failed this cut. Figure 6.22 summarises the effect of the cut on the four data samples tested.

Table 6.9 Event sample sizes prior to and after the impulsive signal cut

	Total	HPOL	VPOL
# Prior to cut	267272	89037	178235
# Removed by flight/traverse	527	148	379
# Removed by impulsive signal cut	1367	132	1235
# Remaining	265378	88757	176621

6.8.1 Effect of Cut

The impulsive signal cut was applied to all events passing thermal cuts that had not already been removed by the flight/traverse clustering cut. It was found that 1367 events failed the cut, the effect on the clustering samples is given in Table 6.9.

The background calculation was updated with the new cut. The background contributions from both thermal and anthropogenic sources were performed in the same manner as described previously. The introduced cut should not increase the thermal background estimate, however, as Figs. 6.23 and 6.24 show, the fraction of thermal noise expected to pass the thermal and impulsive signal cuts now differs from the 2.5×10^{-8} predicted previously. The tails of the distributions of upward-pointing noise events passing the thermal cuts differ very little between Fig. 6.24 and Fig. 6.14. The variation in the fraction of events expected to pass thermal cuts pre- and post-impulsive signal cut can therefore be accounted for by the uncertainty in the original thermal background calculation. Combining the estimates for the number of events passing thermal cuts in Table 6.10, the final thermal background estimates are 0.42 ± 0.24 HPOL and 0.67 ± 0.28 VPOL events. These represent a slight decrease in the HPOL background and a slight increase in the VPOL background. Both estimates are consistent with thermal backgrounds prior to the anthropogenic cut. After event clustering (57 % efficiency, as described previously), these estimates are reduced to 0.24 ± 0.14 HPOL and 0.38 ± 0.16 VPOL events.

The anthropogenic background is calculated using clustering methods that treat events passing all thermal cuts except the requirement on the impulsive nature of signals in two manners; these events can either be left in the clustering sample, but treated as bases, or they can be removed entirely from the clustering sample. By including, or removing, events failing this analysis that passed the analysis in [5], this provides four separate estimates on the updated anthropogenic background. The expected anthropogenic background is 0.07 ± 0.06 HPOL and 0.75 ± 0.22 VPOL (Table 6.11).

The combined background estimate is therefore 0.34 ± 0.15 HPOL events and 1.13 ± 0.27 VPOL events.

The final events passing all thermal cuts, the impulsive signal cut and clustering cuts that were not already removed in the previous section by post unblinding quality cuts are summarised in Table 6.12. Three of the five VPOL candidates passing the analysis in Table 6.8, events # 14917312, # 22807894 and # 23056816, fail the

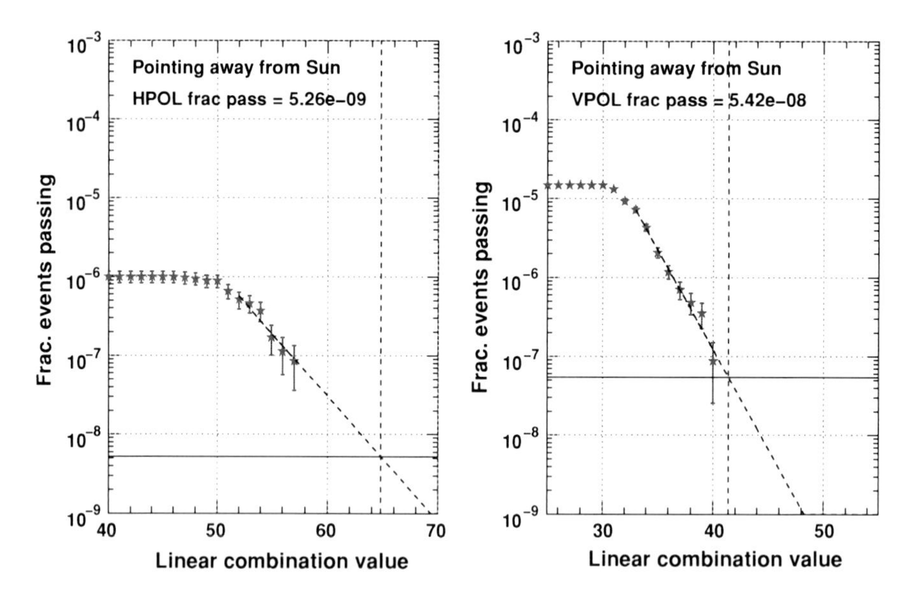

Fig. 6.23 The linear combination cut for HPOL (*left*) and VPOL (*right*) for simulated noise events pointing >20° from the Sun, after impulsive signal cut. Note that the final data points for both plots represent >1 event passing the cut

Table 6.10 Updated thermal background estimates

Sample	HPOL	VPOL
$\Delta\phi_{Sun} > 20°$, before impulsive signal cut	0.5	0.5
$\Delta\phi_{Sun} < 20°$, before impulsive signal cut	0.5	0.5
$\Delta\phi_{Sun} > 20°$, after impulsive signal cut	0.105	1.08
$\Delta\phi_{Sun} < 20°$, after impulsive signal cut	0.792	0.346

Combinations of these for the final thermal estimate take into account $\frac{2}{3}$ of events have $\Delta\phi_{Sun} > 20°$

impulsive signal cut. A total of three HPOL and two VPOL events pass the final analysis cuts, with the event locations displayed in Fig. 6.25.

6.9 Discussion

This chapter has discussed the main analysis on which this thesis is based. The chapter has taken us from a set of 26.7 M events, through event selection cuts, outlined the main analysis tools and demonstrated the cuts used to reject thermal and anthropogenic noise. The analysis has discovered two vertically polarised isolated events and three horizontally polarised isolated events. The horizontally polarised events are considered to be cosmic-ray geosynchrotron emission candidates. Further analysis of these events, and an extension to the cosmic-ray search, are given in

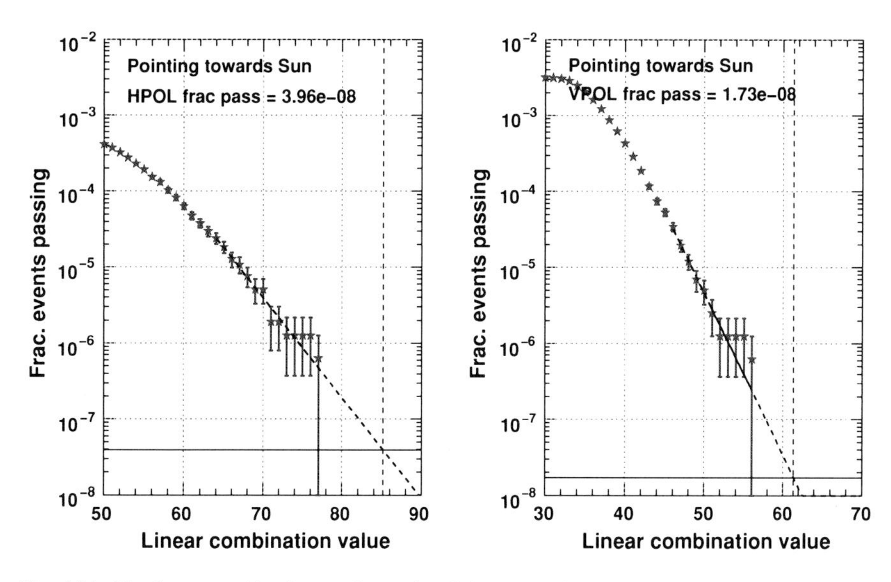

Fig. 6.24 The linear combination cut for HPOL (*left*) and VPOL (*right*) for upward-pointing noise events pointing <20° from the Sun, after impulsive signal cut

Table 6.11 Updated background estimates using four different treatments of the sample of events to cluster

Event sample	Events failing impulsive cut	Base clusters	Base singles	Event clusters	Expected singles
All	Treated as bases	7	1	4	0.571
		2.9 % HPOL, 97.1 % VPOL			0.02 H, 0.56 V
All	Removed	7	1	4	0.571
		2.9 % HPOL, 97.1 % VPOL			0.02 H, 0.56 V
Passing this analysis	Treated as bases	13	4	4	1.23
		10.6 % HPOL, 89.4 % VPOL			0.13 H, 1.10 V
Passing this analysis	Removed	16	5	3	0.938
		14.6 % HPOL, 85.4 % VPOL			0.14 H, 0.80 V

Chap. 7. The vertically polarised events are considered to be neutrino candidates. Further analysis of these events, along with constraints on the diffuse and selected point-source UHE neutrino fluxes, are given in Chap. 8.

The background of 1.13 ± 0.27 VPOL events suggests that the neutrino search discovered no statistical evidence of neutrino-induced radio emission. Moreover, the

Table 6.12 A list of events passing all analysis cuts, including the impulsive signal cut

Event #	Polarisation
14496361	HPOL
14577785	HPOL
15636066	VPOL
16014510	VPOL
27146983	HPOL

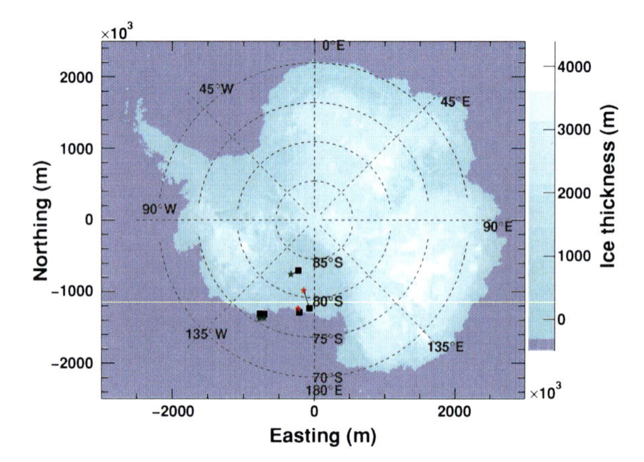

Fig. 6.25 Locations of candidates passing all analysis cuts, *black squares* indicate the balloon position with lines connecting the balloon and event locations, *red stars* indicate VPOL events, *green stars* indicate HPOL events

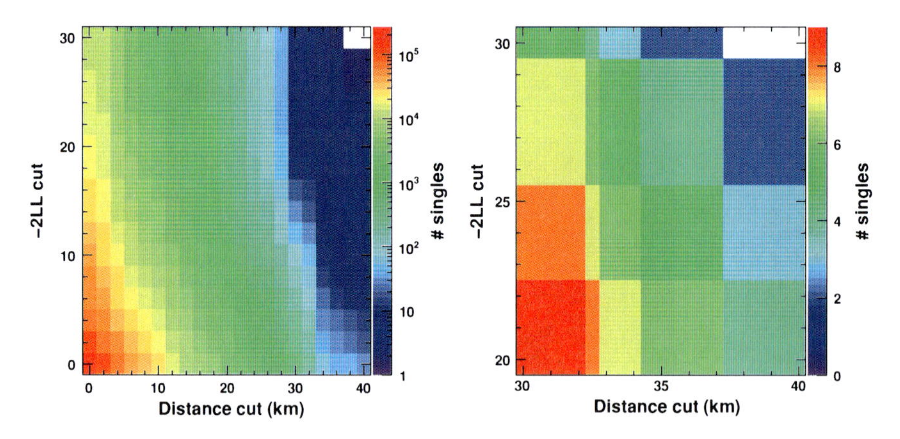

Fig. 6.26 Effect of modifying the clustering separation and $-2LL$ cuts on the number of events passing the clustering. The plot on the *right* is a smaller binned version of the plot on the *left* focussing on the region close to the actual cut values used in the analysis

backgrounds in both polarisations are dominated by anthropogenic noise rather than thermal noise. To test the cuts used in the clustering algorithm, the effect of modifying the cuts on the number of isolated signals was inspected. Figure 6.26 shows that two new events would have been labelled as isolated if either the separation cut were modified from 40 to 37 km, or if the $-2LL$ cut was modified from 30 to 29.

Although this does not confirm the two VPOL candidates as anthropogenic events, the distribution of event singles shown in Fig. 6.26 suggests that the clustering cuts used in the analysis are far from 100 % efficient on removing anthropogenic noise.

References

1. A. Goodhue et al., ANITA internal note, 2008
2. The ANITA Collaboration, P.W. Gorham et al., Phys. Rev. Lett. **103**, 051103 (2009). [astro-ph/0812.2715]
3. S. Hoover, *A Search for Ultrahigh-Energy Neutrinos and Measurement of Cosmic Ray Radio Emission with the Antarctic Impulsive Transient Antenna*. Ph.D. Thesis, University of California, Los Angeles, 2010
4. D. Saltzberg, ANITA internal note, 2009
5. The ANITA Collaboration, P.W. Gorham et al., Phys. Rev. D **82**, 022004 (2010). [astro-ph/1003.2961]

Chapter 7
Cosmic-Ray Search

Analysis of the ANITA-1 data demonstrated the instrument was sensitive to UHECRs [1]. Unfortunately the optimisation of ANITA-2's trigger to UHE neutrinos, which involved running on VPOL signals only, was severely detrimental to ANITA-2's UHECR acceptance.[1] However, the three remaining HPOL signals after all analysis cuts suggest that there may still be UHECR signals in the ANITA-2 data.

7.1 UHECR Search

The UHECR analysis cuts (HPOL) were developed at the same time, and in the same manner, as the neutrino cuts (VPOL). The HPOL event search and clustering was performed in tandem with the VPOL analysis, in this way, a single HPOL and single VPOL signal from the same anthropogenic source which both pass thermal cuts would still be identified and excluded from the final event sample.

As it is possible for ANITA to observe UHECR geosynchrotron emission directly as well as in reflection from the ice, the UHECR search was extended above the horizon (though with $\theta < 0°$, as extensive air-showers will develop below ANITA's flight altitude). Two such events passed all thermal cuts in this energy range; # 20485624 and # 21684774. The first of these is a self-triggered blast event that was not identified by the event quality cuts, the latter event is a well-reconstructed, HPOL dominated event.

The isolated events from the full UHECR search are summarised in Table 7.1. Analysis images of each event are shown in Fig. 7.1, the similarity between the scaled instrument deconvolved coherent waveforms of the four events is demonstrated in Fig. 7.2.

[1] Modifications to the trigger system for ANITA-2 were completed before analysis in [1] demonstrated that ANITA-1 had observed UHECR-induced geosynchrotron radiation.

M. J. Mottram, *A Search for Ultra-High Energy Neutrinos and Cosmic-Rays with ANITA-2*, Springer Theses, DOI: 10.1007/978-3-642-30032-5_7,

Table 7.1 ANITA-2 isolated UHECR candidates

Event #	Elevation (°)	Dir./Refl.	Latitude (°)	Longitude (°)
14496361	−24.9	Reflected	−75.49	−150.73
14577785	−31.2	Reflected	−75.83	−152.779
21684774	−3.0	Direct	N/A	N/A
27146983	−16.1	Reflected	−82.39	−156.39

Latitude and longitude information refers to the reconstructed location on the Antarctic continent rather than projected air-shower location, as such no location information is provided for the direct event

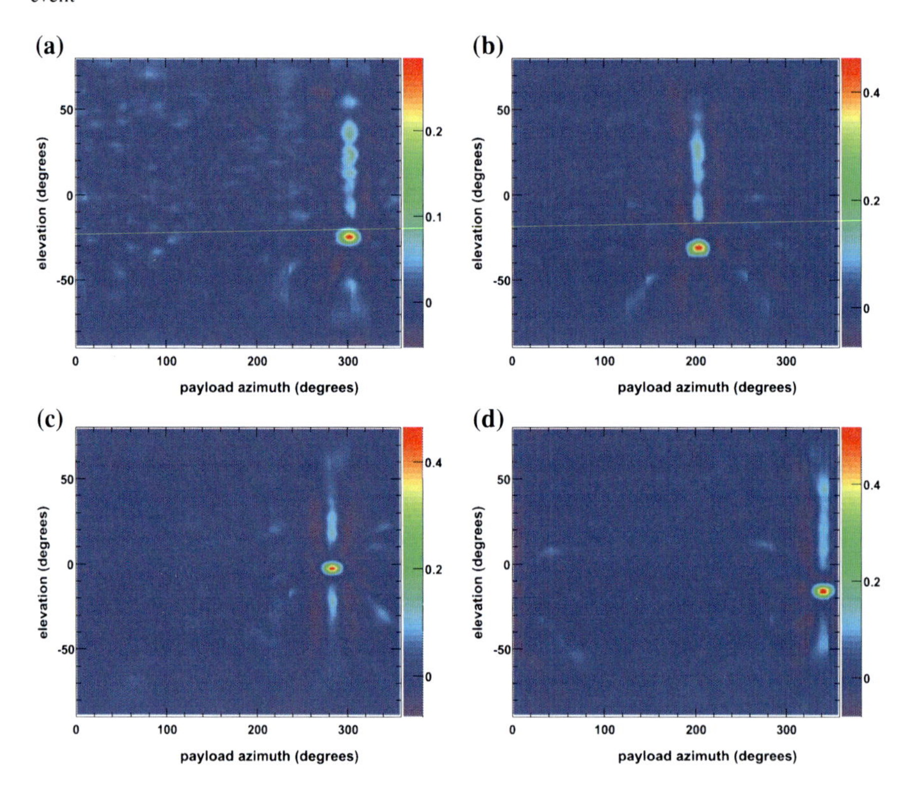

Fig. 7.1 HPOL interferometric images of the four isolated UHECR candidate events, **a**: # 14496361; **b**: # 14577785; **c**: # 21684774; **d**: # 27146983

7.1.1 Identification as UHECR

An UHECR-induced air-shower over the Antarctic continent will develop in a magnetic field, shown in Fig. 7.3, that is predominantly vertical in orientation. The radio emission that ANITA observed from such an air-shower, caused by the splitting and gyration of e^+e^- pairs about the local magnetic field, will therefore be largely horizontally polarised. The UHECR events observed by ANITA-1 [1] were identified

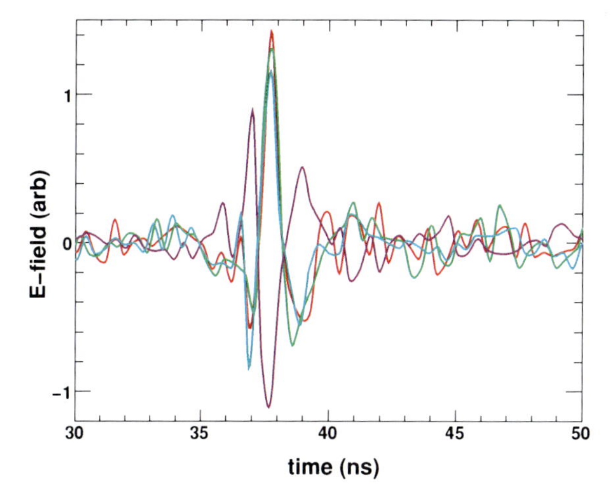

Fig. 7.2 Instrument-response-deconvolved, coherently-summed waveforms for the four isolated UHECR candidates, the directly observed event (*magenta*) displays opposite polarity to the three reflected events. Waveforms have been scaled such that their magnitudes are equal, with a 20 ns time window around the cosmic-ray-induced signal shown (recorded waveforms are ~100 ns long)

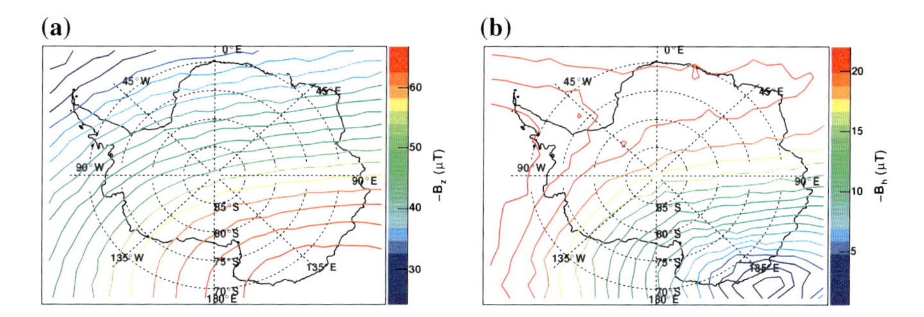

Fig. 7.3 The vertical **a** and horizontal **b** geomagnetic field strength in Antarctica, data from [2]

by comparing the expected polarisation angle of an event with a prediction based on the local magnetic field orientation. Additionally, the power of the geosynchrotron emission was observed to fall exponentially with increasing frequency. This is due to ANITA's measurements being made at frequencies where EAS radio emission has lost coherence, as described in Sect. 3.1.4.

The expected polarisation at emission (θ_{pol}) is found through Eqs. 7.1.1 and 7.1.2, where F_H and F_V are the Lorentz force projected onto local horizontal and vertical axes, $\vec{B}$ is the magnetic field vector at the shower and $\vec{v}$ is the motion of the charge. As the primary cosmic-ray, the longitudinal shower axis and the radio emission are all expected to be fairly collinear, $\vec{v}$ is taken as the projected RF emission direction.

$$\vec{F} = \vec{v} \times \vec{B} \tag{7.1.1}$$

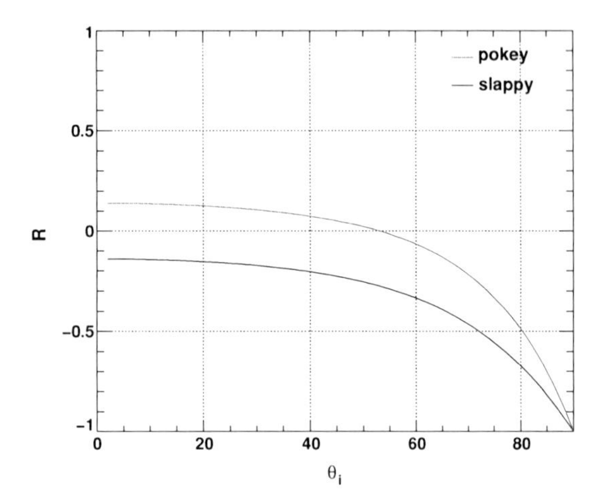

Fig. 7.4 Fresnel coefficient of reflection as a function of incidence angle for VPOL ($E_{\parallel}$) and HPOL ($E_{\perp}$)

Table 7.2 Comparison of measured and expected VPOL signal strengths in the case that all four candidates are caused by UHECR EAS geosynchrotron emission

Event #	HPOL mV/m	VPOL Expected mV/m	VPOL Observed mV/m
14496361	1.10	0.090	0.332
14577785	1.15	0.046	0.355
21684774	1.86	0.399	0.597
27146983	2.67	0.377	0.362

The largest discrepancy occurs for highly down-pointing events (i.e. events with large negative elevations), a detector effect is the likely cause. Measured signal strengths are taken as the electric field of the three ϕ-sector instrument deconvolved coherent waveforms

$$\theta_{pol} = \tan^{-1}\left(\frac{F_V}{F_H}\right) \tag{7.1.2}$$

Using the magnetic field orientation at projected shower location for each event, estimates of the expected VPOL signal strength at the balloon were made. These were based on the observed HPOL signal strength and accounted for Fresnel reflection coefficients, shown in Fig. 7.4, where appropriate. Table 7.2 shows that the two highly inclined showers displayed an excess in VPOL signals when compared to the expectation. Upon inspecting these events, the VPOL signal appears almost entirely in antennas on either side of the main ϕ-sector in which HPOL emission was observed. This implies that the mismatch between expected and observed VPOL signals (and, hence, event polarisation) could be due to a detector effect.

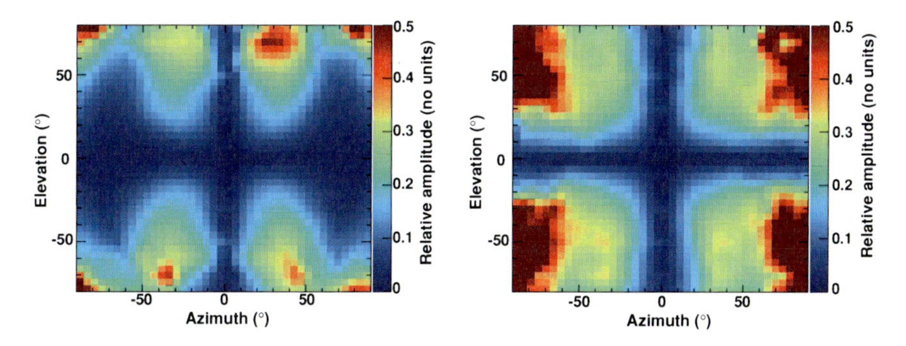

Fig. 7.5 Relative peak-peak amplitude of the cross-polarisation response as a function of angle off-boresight for HPOL→VPOL (*left*) and VPOL→HPOL (*right*)

7.1.2 Antenna Response

Until now, all analysis has neglected the effect of signal leakage at the antenna stage between the two ANITA polarisations. Calibration signals from the Taylor Dome borehole antenna were transmitted only in VPOL. It was observed in the ANITA-2 data that, when the balloon was very close to Taylor Dome, these signals were also observed in HPOL. It was noted that the HPOL signal was often stronger in antennas further off-boresight in azimuth than those on boresight. It became apparent that the observed HPOL signals were caused by non-negligible cross-polarisation effects in the antenna. These effects are known to become increasingly significant the further off antenna boresight a signal is. This effect could easily impact on any physics signal observed in the ANITA-2 data, particularly as both UHECR EAS emission and neutrino-induced Askaryan emission that ANITA-2 was sensitive to are expected to be highly linearly polarised.

Prior to the first flight of ANITA, a series of tests was conducted on the Seavey antennas in an anechoic chamber at the University of Hawai'i [3]. Included in this data are the response of antennas to the transmission of impulsive signals, with transmissions made in either H- or V-POL and response measured for both polarisations for every measurement. The data was taken over a range of angles for seven different antennas, it is this data that is used in the system response for both simulation and signal deconvolution. However, perpendicular-polarised response measurements were only made in all seven antennas for one angle (azimuth or elevation) being off-boresight at a time. Figure 7.5 shows the results of cross-polarisation measurements, made for one Seavey antenna in [3] that include off-boresight response in both azimuth and elevation angles.

Taylor Dome Results

Using Taylor Dome impulsive events, the cross-polarisation antenna response was tested by comparing a prediction of the HPOL signal (induced by VPOL emission) with the measured HPOL signal in each event. Even within the VPOL measurements, there are significant fluctuations in signal strength at an event-to-event level. As the HPOL signal prediction is made for each antenna and each event, the resulting HPOL prediction also varies greatly from event to event.

The cross-polarisation leakage from VPOL to HPOL is only well defined for angles of incidence up to $\sim 45^{\circ}$ off-boresight. While the main analysis uses five ϕ-sectors of antennas to produce coherently-summed waveforms, the amplitudes of predicted HPOL were too uncertain in the outermost ϕ-sectors for the Taylor Dome events. The coherently-summed waveforms used in this section were therefore restricted to three payload ϕ-sectors, corresponding to seven or eight antennas (depending on the nadir antenna locations).

Figure 7.6 demonstrates that the HPOL signal amplitude prediction follows the observed HPOL amplitude closely. However, Fig. 7.7 shows that the predicted amplitude of the coherently-summed waveform from cross-polarisation leakage is consistently lower than the observed amplitude. It is possible that the discrepancy is caused by time delay effects introduced in the cross-polarisation response that depend on angle of incidence. Improving the prediction would be difficult; as well as having only cross polarisation response data from one antenna, the Taylor Dome calibration signals often contain CW contamination which affects the measured HPOL amplitude.

7.1.3 Isolated Event Results

Using the expected cross-polarisation response effects, the measured and expected polarisation angles of the four UHECR candidate events are once again compared. Figures 7.8 and 7.9 demonstrate that, with the observed VPOL E-field amplitudes closely matching the expected amplitudes, the resulting expected and observed polarisation angles are therefore also well matched.

7.1.4 Non-Isolated Events

The anthropogenic event cuts described in Sect. 6.5 removed any ANITA-2 events that clustered with one or more other non-thermal events. Within this clustering analysis, three event doublets were found—that is, three clusters consisting of two events and no bases. In two of these doublets, both events passed the thermal cuts described in this analysis. The third doublet contained only one event passing this analysis, with the other event coming from the inserted events passing the analysis described in [4].

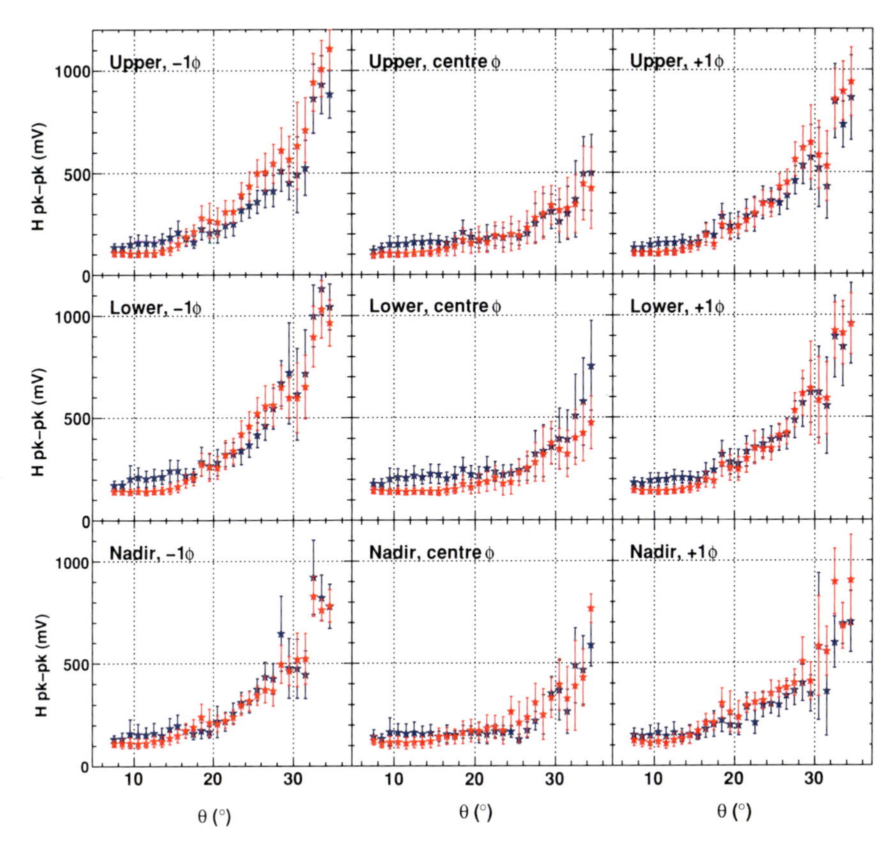

Fig. 7.6 Predicted waveform amplitudes (*red*) and measured HPOL waveform amplitudes as a function of elevation angle for Taylor Dome events in three sectors of antennas

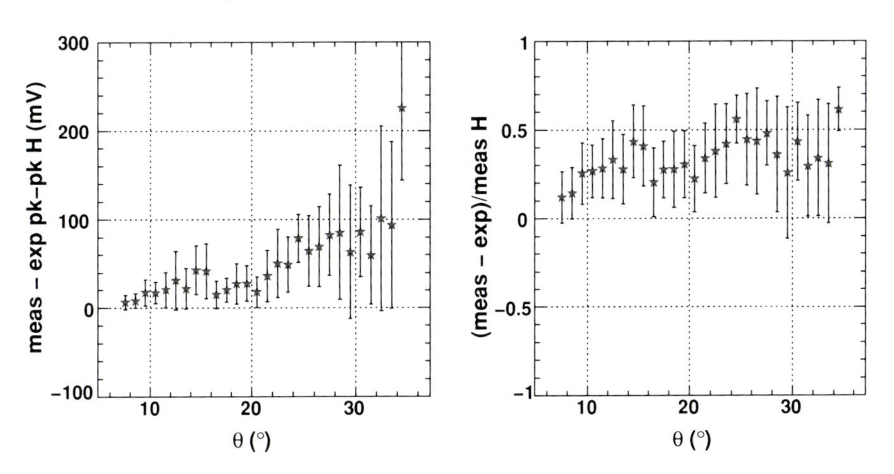

Fig. 7.7 Absolute (*left*) and fractional (*right*) error in HPOL prediction from cross polarisation signals for coherently-summed waveforms from Taylor Dome as a function of elevation angle

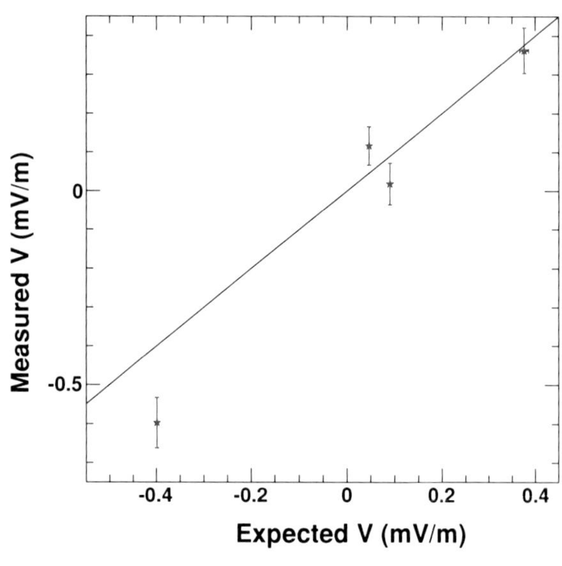

Fig. 7.8 Measured versus expected VPOL E-field amplitudes of UHECR emission for the four isolated cosmic-ray candidates. Expected values include Fresnel coefficients, measured values include cross-polarisation signal contamination. The *black line* is measured = expected

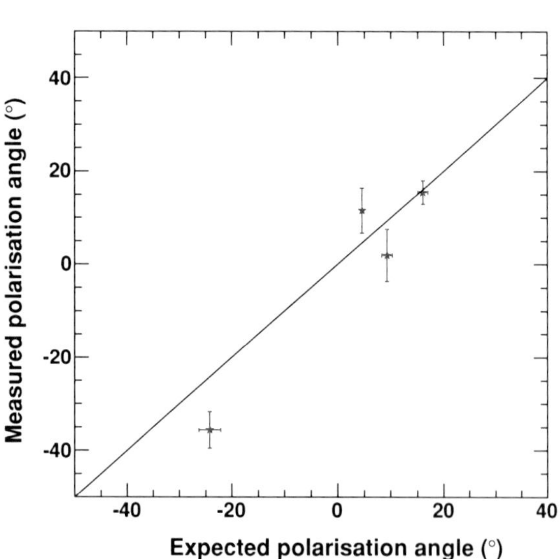

Fig. 7.9 Measured versus expected polarisation angles of UHECR emission for the four isolated cosmic-ray candidates. Expected values includes Fresnel coefficients, measured values include cross-polarisation signal contamination. The *black line* is measured = expected

The event failing this analysis contained narrowband CW that was filtered. The remaining event, # 3493989, passed the analysis in HPOL and reconstructed with an elevation of $-30°$.

To search for further UHECR events in the ANITA-2 data, the coherent waveforms of all events passing thermal cuts are compared to the coherent waveforms of the isolated (reflected) UHECR candidates. Both the instrument-response-included and instrument-response-deconvolved, coherently-summed, waveforms were compared

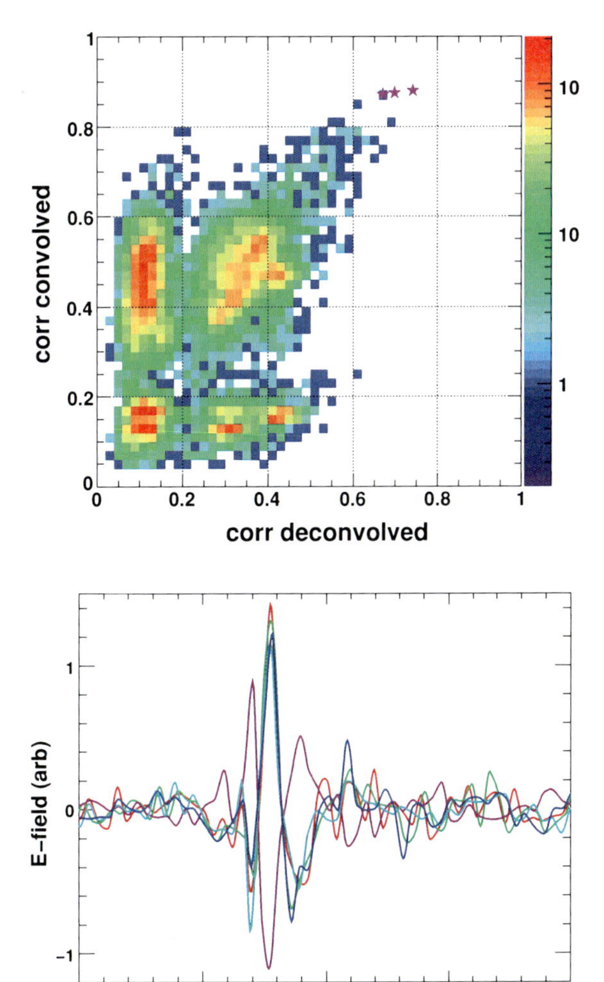

Fig. 7.10 Average correlation coefficients of waveforms for all events passing the HPOL thermal cuts to the waveforms of the three isolated, reflected, UHECR candidates. Correlation coefficients are found for both the normalised instrument-response-deconvolved (x-axis) and normalised instrument-response-included (y-axis) coherently-summed waveforms. *Magenta stars* indicate the average correlation of each UHECR candidate to the other two UHECR candidates. The correlation values for event # 3493989 are 0.67, 0.87. Note that events with negative correlation values are not included in the plot

Fig. 7.11 Waveforms from Fig. 7.2 with the one non-isolated candidate event overlaid in *dark blue*

by correlating the UHECR candidate waveforms with each event's HPOL waveforms. Average correlation coefficients to the three candidates were calculated for each event and compared to the average correlation of the candidates with each of the other two candidates. The results of this test are shown in Fig. 7.10. It was found that event # 3493989 correlates with the reflected isolated UHECR events very well compared to all other HPOL events that had been cut by event clustering. In fact, the average correlation between event # 3493989 and three reflected UHECR candidates is as good as the correlation of cosmic-ray event # 14496361 to the remaining two candidates.

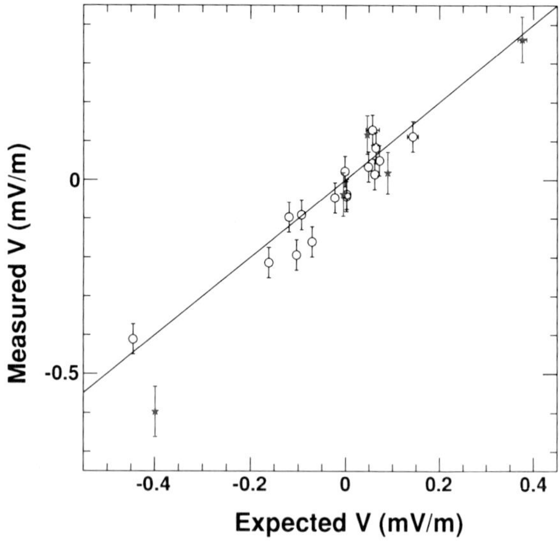

Fig. 7.12 Measured versus expected VPOL E-field amplitudes of UHECR emission for the four isolated and one non-isolated cosmic-ray candidates (*red stars*). Expected values include Fresnel coefficients, measured values include cross-polarisation signal contamination. The 16 UHECR events from ANITA-1 are also displayed (*black circles*). The *black line* is measured = expected

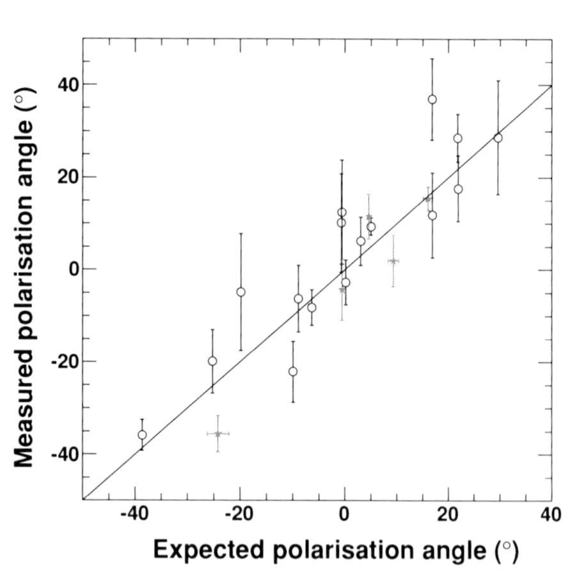

Fig. 7.13 Measured versus expected polarisation angles of UHECR emission for the four isolated and one non-isolated cosmic-ray candidates (*red stars*). Expected values include Fresnel coefficients, measured values include cross-polarisation signal contamination. The 16 UHECR events from ANITA-1 are also displayed (*black circles*). The *black line* is measured = expected

The waveforms for the isolated UHECR candidates and event # 3493989 are shown in Fig. 7.11. Figures 7.12 and 7.13 show that the expected and observed VPOL emission were also well matched. It is highly likely that event # 3493989 was caused by geosynchrotron emission from an UHECR-induced EAS.

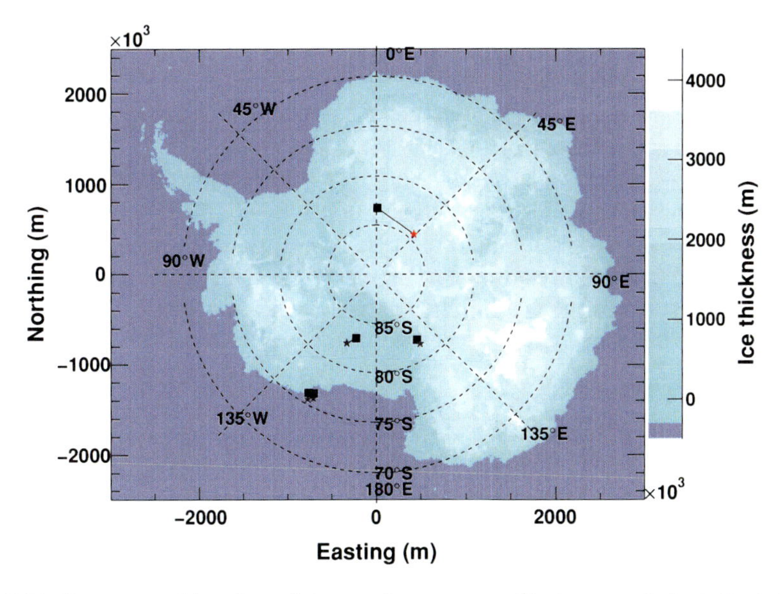

Fig. 7.14 Reconstructed locations of the cosmic-ray events. *Black squares* indicate the balloon location. The three reflected events from the main analysis are shown with *green stars*, the location of closest approach to ground of the direct event is indicated by the *red star*, the non-isolated event is indicated with a *blue star*

7.2 Discussion

ANITA-2 observed four isolated, horizontally polarised events that have been shown to be signals from geosynchrotron radiation of cosmic-ray-induced extensive air-showers. The data also show a further event that, while clustered with one other reconstructed event, is highly likely to be a signal from UHECR geosynchrotron emission. The locations of the events found in the ANITA-2 cosmic-ray search are shown in Fig. 7.14.

The ANITA-1 data analysis discussed in [1] observed 16 isolated UHECR events. The uncertainty in the energy calculation for these events was relatively large, however, simulation- and data-driven estimations for this energy calculation place the mean primary energy from the 16 events at $>10^{19}$ eV. The ANITA-1 results represented the first observations of UHECR-induced geosynchrotron emission at frequencies $>$600 MHz.

Although the ANITA-2 data only provides a further four isolated UHECR (five events if we include the non-isolated correlated event), it does represent an opportunity to produce an independent measurement of the frequency spectra of UHECR geosynchrotron.

7.2.1 Frequency Dependence

The power of geosynchrotron emission from UHECR-induced air-showers is expected to fall off rapidly with increasing frequency for wavelengths shorter than the transverse shower size. At such wavelengths, the substructure of the shower is being resolved, so emission is no longer coherent over the shower. Electromagnetic shower transverse size is governed by the Moliére radius. In air, geosynchrotron emission will begin to lose coherence at frequencies of $O(10)$ MHz (exact values depend on the size of the shower). Spectral emission measurements from the ANITA-1 UHECR events confirm this frequency dependence, with a spectrum of

$$I(\nu) \propto e^{-\left(\frac{\nu}{121\,\mathrm{MHz}}\right)} \mathrm{W\,m^{-2}\,MHz^{-1}} \tag{7.2.1}$$

observed for the coherently-summed average of the 16 observed cosmic-ray events.

The frequency spectrum was calculated for the four isolated UHECR events (omitting the clustered candidate to retain a conservative analysis). The power of emission from the individual UHECR events that ANITA-2 observed falls below the thermal noise floor for frequencies above about 600 MHz. To assist with resolving power at higher frequencies, a 10 ns window is taken around the peak of the instrument-response-deconvolved, coherently-summed, waveforms. Electric field values at all other times are set to zero to remove thermal noise. The four individual signals are then normalised in amplitude and coherently-summed. The coherent power spectrum for the four event average is shown in Fig. 7.15.

An exponential function appears to fit the coherently averaged frequency spectrum of the cosmic-ray events. The exponential function fitted in Fig. 7.15 follows

$$I(\nu) \propto e^{-\left(\frac{\nu}{376\,\mathrm{MHz}}\right)} \mathrm{W\,m^{-2}\,MHz^{-1}} \tag{7.2.2}$$

This represents a harder spectrum than was found for the ANITA-1 events.

Thermal noise levels for the averaged coherent power spectrum ($10^{-6} - 10^{-5}$ in the arbitrary units of Fig. 7.15) suggest that thermal fluctuations are not the cause of the discrepancy between the ANITA-1 and ANITA-2 power spectrum measurements. It is possible that the different signal chain responses (and the level of accuracy to which they were measured) could contribute to the observerd difference. Another potential cause is the trigger banding and logic, which differed between ANITA-1 and ANITA-2. In ANITA-1, two circularly polarised channels (LCP and RCP) were created by combining HPOL and VPOL signals, with each channel then divided into four frequency bands for the sub-band trigger. The four bands (low, mid1, mid2 and high) roughly covered 200–400, 400–600, 600–800 and 800–1,200 MHz respectively, only the low band was retained in ANITA-2. There was a 3-of-8 L1 trigger requirement for an antenna-wide trigger, such that a signal could be recorded with just two low band triggers plus one trigger from another band. In addition to a reduced aperture to cosmic-rays by running on a VPOL-only trigger, the trigger banding used in ANITA-2 of 2-of-3 low, mid or high-band in addition to a required full-band

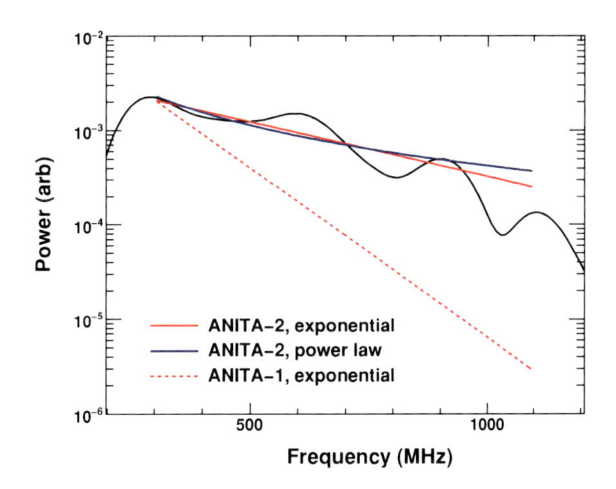

Fig. 7.15 The coherent power spectrum of the four cosmic-ray events, with arbitrary power units. Two fits to the data in a $300 < f < 1,000$ MHz range are displayed, a power law and exponential. The exponential best-fit to ANITA-1 data is also shown, having been scaled such that the power at 300 MHz matches that of the ANITA-2 exponential best-fit. The thermal noise power level ranges from 10^{-6} to 10^{-5} in the frequency range of interest, these values are negligible compared to the power of the cosmic ray signal

trigger biased ANITA-2 to recording cosmic-ray signals with increased high-frequency content.

7.2.2 Event Energies and Directions

The UHECRs observed by ANITA-1 are likely to be highest energy events observed via geosynchrotron emission to date. The energy scale is still uncertain, though the mean energy of the 16 observed events is expected to be $>10^{19}$ eV. The events observed by ANITA-2 are likely to have similar energies, though the two isolated and one non-isolated highly down-pointing events will be at the low energy end of this distribution. Table 7.3 gives the electric field strength at the shower for each candidate. The average electric field strength of the ANITA-1 events was found to be 325 V/m. The three reflected events that originated very close to the balloon have lower field strengths than this value, as expected. Assuming that all of the cosmic-rays are viewed in the same manner, the energy of the events scale linearly with the electric field strength of the emission, implying the mean energy of the ANITA-2 cosmic-rays is similar to that of the ANITA-1 events. It should be noted, however, that this assumption is not likely to be a reliable one—the frequency spectrum of EAS geosynchrotron emission is expected to vary significantly with the angle from which observations are made relative to the primary cosmic-ray's direction. We known that the frequency spectra of the events observed by ANITA-2 differs from those observed by ANITA-1. Unfortunately, modelling of geosynchrotron at such high energies in the far field scenario remains inconsistent with the ANITA observations and an estimation of viewing angle is not possible.

Table 7.3 Electric field at the projected location of the UHECR EAS

Event #	Projected distance (km)	r	E-field at 1 m (V/m)
3493989	76	−0.319	225 ± 21
14496361	107	−0.388	249 ± 58
14577785	85	−0.325	317 ± 38
21684774	505	N/A	1095 ± 183*
27146983	154	−0.542	573 ± 3

Distance values are given as 1.2 times the distance to site of reflection for reflected events (resulting in an estimated altitude of shower maximum of 7 km) and as the distance to location of closest approach for the direct event. While Fresnel coefficients of reflection, r, are accounted for, surface roughness effects are not considered. The effect of surface roughness for the three events from the Ross Ice Shelf or sea ice should be negligible. The uncertainties given only account for variation between signal strengths from different antennas

* The direct event has a projected shower maximum very far from the balloon, if the emission was viewed slightly ($<2°$) off shower axis then the projected location could be significantly closer to the balloon (as little as 200 km), resulting in a significantly reduced projected E-field.

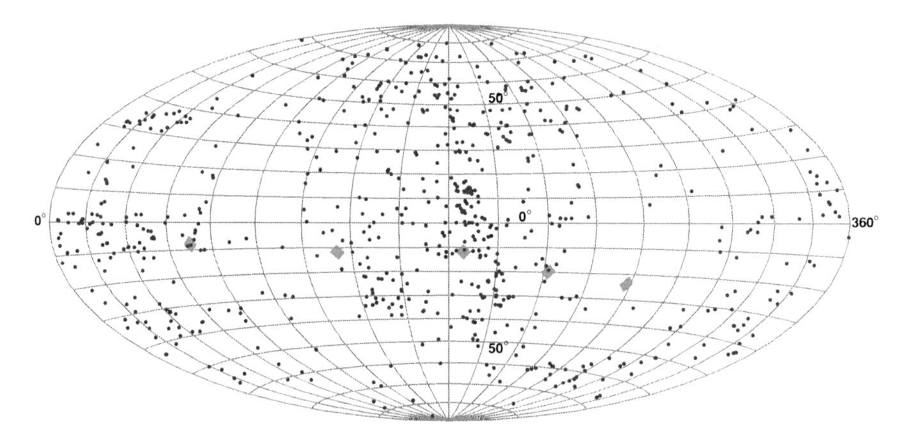

Fig. 7.16 Projection of the four isolated and one non-isolated cosmic-ray candidates in *right* ascension and declination. Errors on the projection are expected to be $\leq 5°$. *Black dots* represent AGN within 100 Mpc from the Veron-Cetty catalogue [5]

Cosmic-rays at energies $E < 10^{20}$ eV are not expected to reconstruct back to their source location. However, we can still reconstruct the UHECR primary arrival direction in celestial coordinates as shown in Fig. 7.16.

7.2.3 Outlook for ANITA-3

The confirmation of ANITA-1's observation of geosynchrotron emission from extensive air-showers was provided after the ANITA-2 flight. Through attempts to optimise the hardware sensitivity to neutrino-induced signals, ANITA-2's UHECR acceptance was severely reduced when compared with ANITA-1. A third flight of the ANITA experiment, ANITA-3, will provide an opportunity to implement the noise reducing modifications that were used for ANITA-2's VPOL channels in the HPOL channels as well. Linearly polarised triggers will still be used, but with dedicated and independent horizontal and vertical polarisations. As such, ANITA-3 is expected to have a vastly increased aperture to UHECR signals when compared to both of its predecessors. Estimates of the UHECR sample size for ANITA-3 are of $O(100)$ events for a typical 30 day flight, an order of magnitude increase over the combined ANITA-1 and ANITA-2 data. The majority of this data will consist of events at energies lower than the mean ANITA-1 energy, though it is expected that ANITA-3 will be able to detect a larger sample of events with $E > 10^{19}$ eV than both of its predecessors.

References

1. The ANITA Collaboration, S. Hoover et al., Phys. Rev. Lett. **105**, 151101 (2010), [astro-ph/1005.0035]
2. S. McLean et al., NOAA Technical Report NESDIS/NGDC-1 (2004)
3. P. Miocinovic, ANITA internal note, 2006
4. The ANITA Collaboration, P.W. Gorham et al., Phys. Rev. **D82**, 022004 (2010), [astro-ph/1003.2961]
5. M. Véron-Cetty, P. Véron, Astron. Astrophys. **518**, A10+ (2010)

Chapter 8
Neutrino Search

The search for isolated VPOL RF signals within the ANITA-2 data returned two events that the analysis classed as isolated signals consistent with the expectation from neutrino interactions. Using the expected number of background events, this result can be used to place a constraint on the flux of UHE neutrinos. Such a limit is implemented in this chapter in two ways. Firstly, it is possible to place an overall, model-independent, limit on the UHE neutrino flux as a function of primary energy. Secondly, ANITA-2's directionally dependent exposure is used to place UHE neutrino flux limits on specific sources.

8.1 Results of the Main Analysis

The analysis outlined in Chap. 6 returned two vertically polarised events that passed all thermal and clustering cuts. Using this result together with the aperture of ANITA-2 (see Fig. 5.10), the analysis efficiency (see Fig. 6.20) and the expected background (see Sects. 6.6.2 and 6.8), it is possible to set a limit on the diffuse UHE neutrino flux.

8.1.1 Neutrino Candidates

Event 15636066

The interferometric image and coherently-summed, instrument-response deconvolved, waveform for event # 15636066 are shown in Fig. 8.1. The event is highly down-pointing, with $\theta \sim -34\,^\circ$, so only just passes the elevation cut. The event was recorded when ϕ-sector masking was activated in ϕ-sectors 1 & 2, CW is present in the spectra of the masked channels. The trigger originated in ϕ–sectors 5 & 6,

M. J. Mottram, *A Search for Ultra-High Energy Neutrinos and Cosmic-Rays with ANITA-2*, Springer Theses, DOI: 10.1007/978-3-642-30032-5_8,

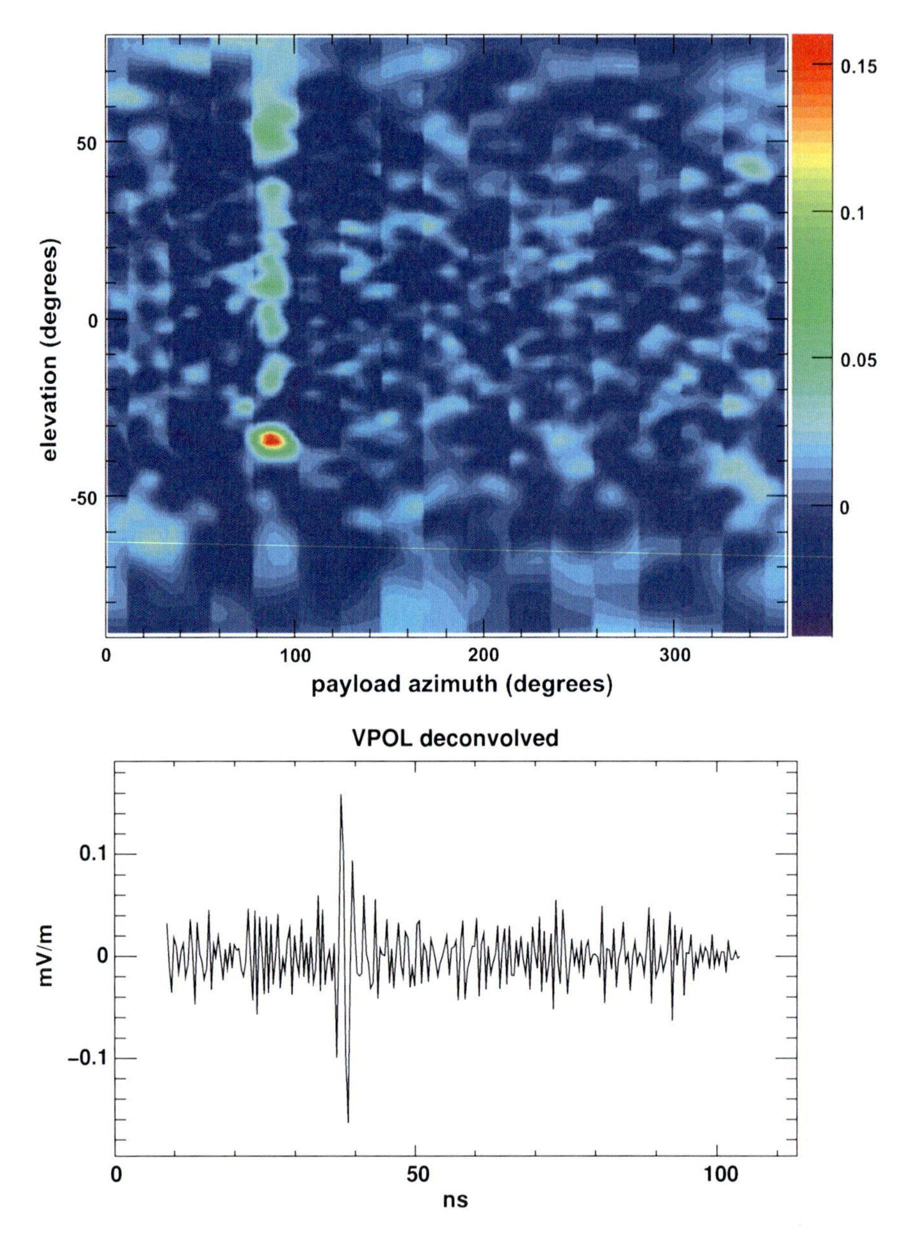

Fig. 8.1 Event #15636066. *Top* interferometric image, *bottom* coherently-summed, instrument-response-deconvolved, waveform

over 45 ° from the masked channels, suggesting the source of the triggered signal and CW are not associated. The CW noise was successfully removed by the filtering algorithm and the event points in the direction of the triggered channels. The event reconstructs to a location close to the edge of the Ross Ice Shelf. The closest base and event passing thermal cuts to event # 15636066 are 51.3 and 46.9 km away respectively (each with $-2LL > 1500$).

The instrument deconvolved waveform of the event displays a strong impulsive event that appears bipolar in nature. The horizontally polarised channels display no similar signals and the event has a polarisation of 86° from the horizontal after filtering.

Event 16014510

The interferometric image and coherently-summed, instrument-response-deconvolved, waveform for event # 15636066 are shown in Fig. 8.2. Event #16014510 is a well reconstructed event, with a peak correlation coefficient of ~0.23. The event originates from the Ross Ice Shelf. The closest base and event passing thermal cuts to event # 16014510 are 149.6 and 41.3 km away respectively, with $-2LL$ of 198 and 132.

The instrument deconvolved waveform is less clearly impulsive than that of event #15636066, event #16014510 is also slightly less VPOL dominated with a polarisation of 83° from the horizontal. Neither of these characteristics are inconsistent with expected signals from neutrino induced EM showers.

Event Locations

Both isolated VPOL events originate from the Ross Ice Shelf, from locations close to clusters of anthropogenic noise. Simulations inform us that ANITA-2 is far more likely to observe neutrino induced radio signals from regions of >km deep ice, rather than the shallow ice shelves, as shown in Fig. 8.3. Figure 8.4 compares the clustering results for the candidate events to those for simulated neutrinos. Both events only just passed the distance cut, which defined events as isolated if their separation from bases and other events was >40 km. It therefore seems highly likely that the two events passing all analysis cuts are background events of anthropogenic origin.

As both events were classed as isolated by the clustering algorithm they are considered neutrino candidates in order to reconstruct the direction of the potential neutrino primary. This direction is then mapped onto celestial coordinates in Fig. 8.5. Event # 16014510 reconstructs with a neutrino primary direction of right ascension $\alpha = 193.5°$ and declination $\delta = -9.1°$. Event # 15636066 reconstructs with a neutrino primary direction of $\alpha = 80.4°$ and $\delta = -15.5°$. The latter of these locations is at the very edge of ANITA-2's declination exposure to neutrinos in the case of a Standard Model cross section (see Sect. 8.2.1).

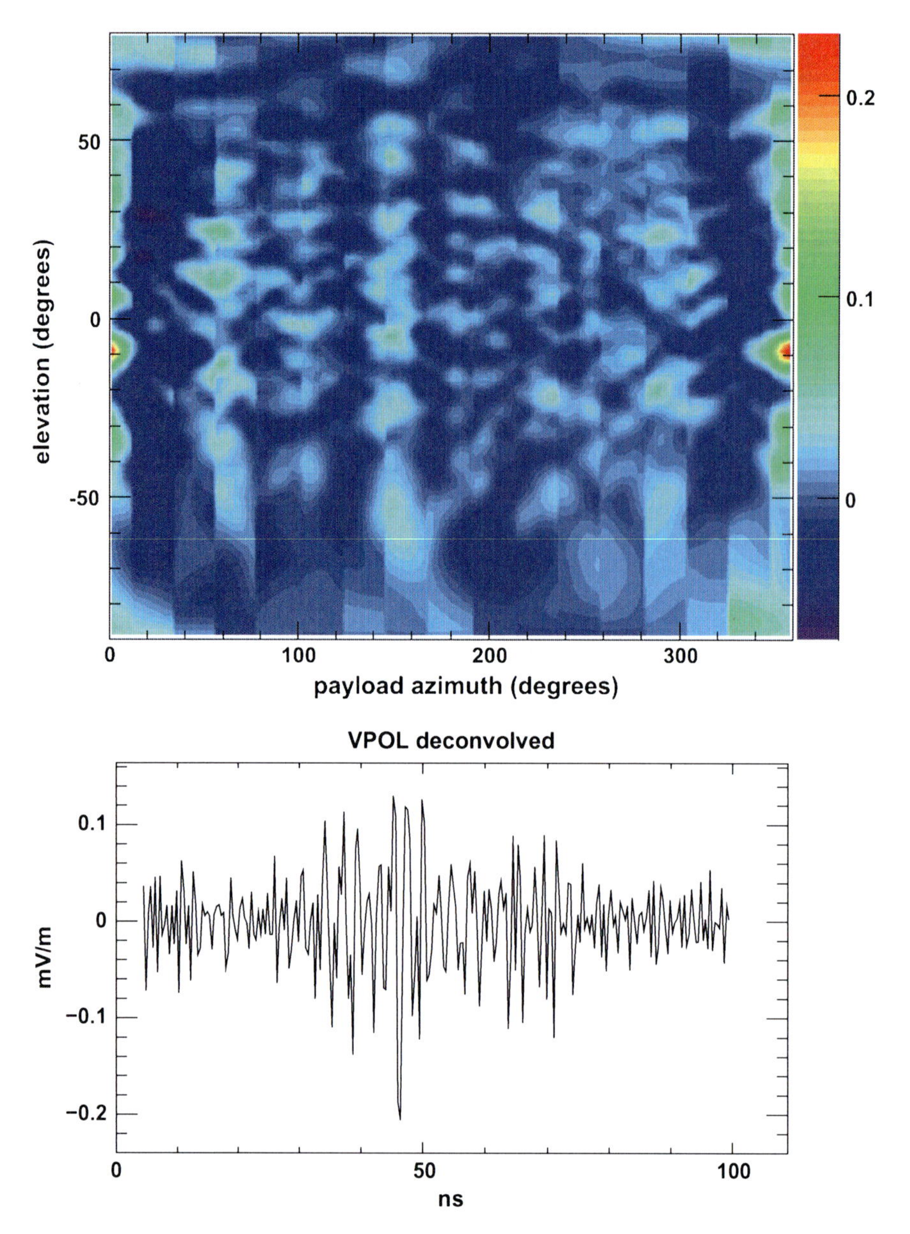

Fig. 8.2 Event #16014510. *Top* interferometric image, *bottom* coherent-summed, instrument-response-deconvolved, waveform

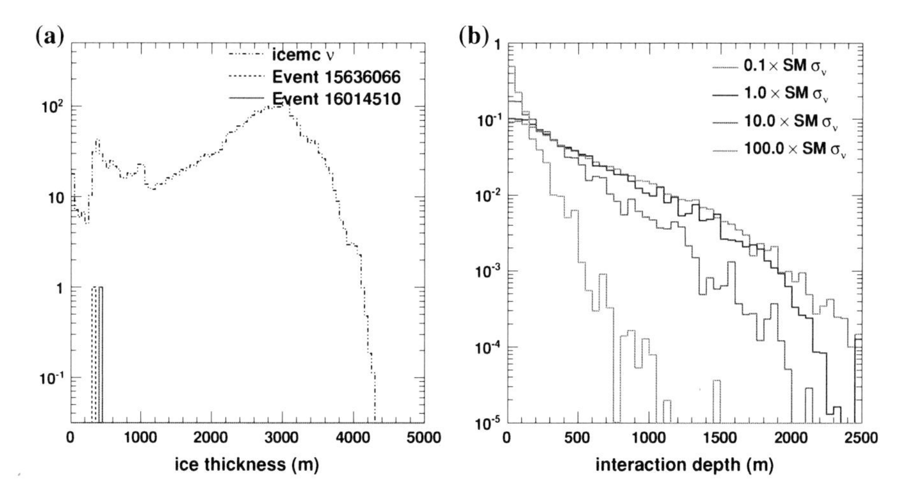

Fig. 8.3 **a** The ice thickness at reconstructed event location for the two neutrino candidates and simulated neutrino events. **b** The effect of the extrapolated neutrino cross section (from [1]) on the average depth of interaction of simulated neutrinos that cause a hardware trigger. Histograms shown in (**b**) have been normalised

8.1.2 Diffuse UHE Neutrino Flux Limit

Using the neutrino search results a limit can be placed on the UHE neutrino flux. Naively, it can be assumed that the upper limit on the flux can be defined via Eq. 8.1.1.

$$E_\nu \frac{dN}{dE_\nu} \leq \frac{N_{90}}{A_{\mathrm{eff}}(E_\nu)T_{\mathrm{live}}(E_\nu)\varepsilon(E_\nu)} \tag{8.1.1}$$

Where N_{90} is the 90 % upper confidence limit on the number of neutrino events excluded by the analysis, A_{eff} is the aperture of the experiment in km^2 sr, T_{live} is the livetime of the experiment and ε is the efficiency of the analysis. A_{eff}, T_{live} and ε are all energy dependent.

Methods set out in [2–4] show that Eq. 8.1.1 underestimates the limit that can be placed on the UHE neutrino flux. Equation 8.1.1 effectively places an upper limit on the flux at any given neutrino energy for the number of events observed, while the desired limit should account for neutrino flux models providing neutrinos over the energy interval being considered. [2–4] argue that both neutrino flux and experimental exposure can be represented as a piecewise construction of power laws. Using slightly different approaches, the arguments of both [2, 4] and [3] show that the limit calculated in Eq. 8.1.1 should be scaled by a factor of $\frac{1}{4}$. This scaling is applied to the limit and is consistent with the scaling used in [5].

Two different A_{eff} values are used, from the *icemc* Monte-Carlo simulation and the modified *icemcEventMaker* simulation (described in Chap. 5 and shown in Fig. 5.10). T_{live} is 28.5 days and ε is taken from Fig. 6.20. Feldman Cousin's statistics [6] are used

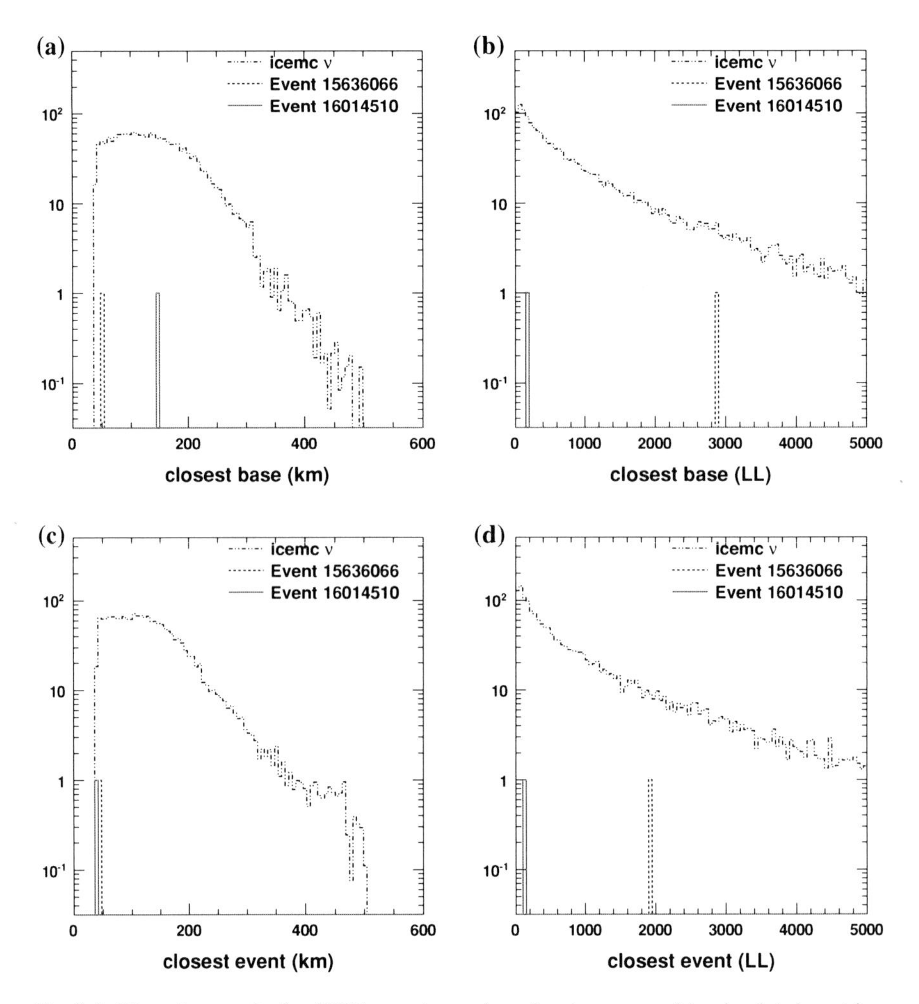

Fig. 8.4 Clustering results for VPOL events passing all cuts compared to simulated neutrinos passing all cuts. **a** Closest base separation. **b** Closest base log-likelihood. **c** Closest event separation. **d** Closest event log-likelihood

to calculate N_{90}; the two observed events on an expected background of 1.13 ± 0.27 lead to $N_{90} = 4.785$.

The resultant UHE neutrino flux limit is shown in Fig. 8.6. The flux limit calculated in this analysis is not an improvement over that calculated in [12], where one VPOL event was observed on a background of approximately one (see [5] for revised result). A number of factors contribute to the limit placed here being less stringent. Two events were observed on an expected background of 1.13 ± 0.27, resulting in a higher value of N_{90}. This analysis included events that failed thermal cuts, but passed the thermal cuts of [12], at the clustering stage. This resulted in a reduced area in which an event would be allowed to pass the clustering cuts. Finally, the analysis described in [12]

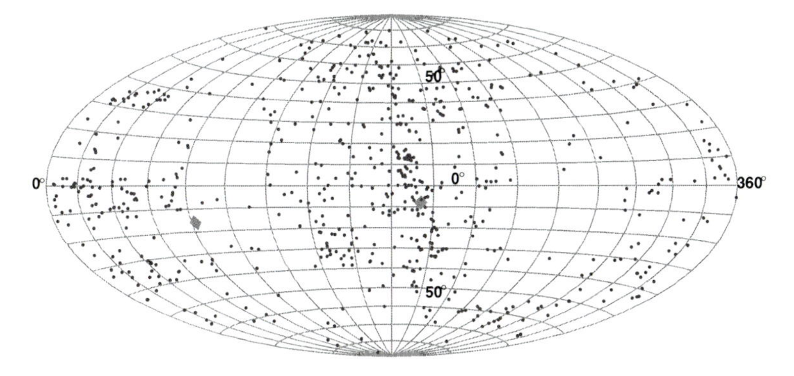

Fig. 8.5 Projection of isolated VPOL candidates in right ascension and declination in the case that the observed signals are Askaryan pulses from neutrino induced EM showers. Note that the size of the projected areas are not representative of the neutrino direction resolution. Black dots represent AGN within 100 Mpc from the Veron-Cetty catalogue

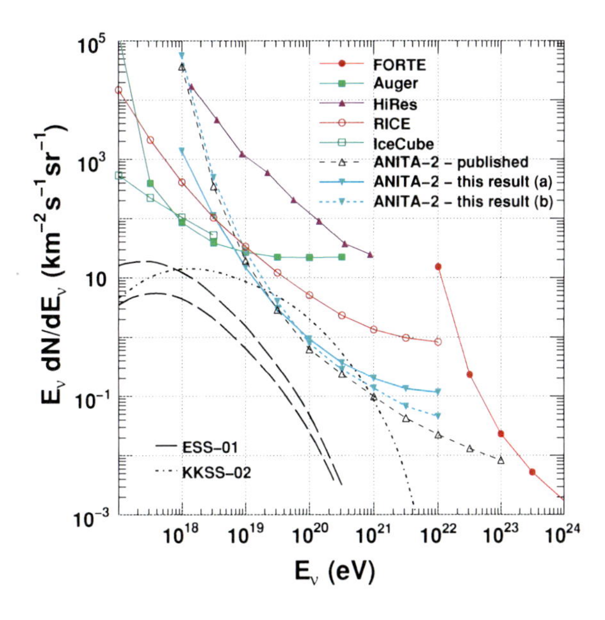

Fig. 8.6 Limit on the diffuse neutrino flux from the analysis described in this thesis. Two limits are shown for the ANITA-2 analysis from this work, using the aperture from the *icemcEventMaker* modified simulation (**a**) and the aperture from the *icemc* simulation (**b**) as calculated in Chap. 5. Also shown are limits from the FORTE [2], Auger [7], HiRes [8], RICE [3] and IceCube [9] experiments, as well as the published neutrino limit for ANITA-2 [5]. ESS-01 lines show the flux of ν_μ for two source evolution scenarios from [10]. The KKSS-02 line shows the flux predicted by [11] in the case of strong source evolution and an injected UHECR spectrum that follows $E_{CR} \propto E^{-1}$

did not analyse simulated neutrinos and, as such, did not account for events that would be cut due to SURF saturation.

If the five VPOL events found prior to the impulsive signal cut are used, with an expected background of 1.5 events, $N_{90} = 8.49$. The resultant limit would be a factor of 1.8 weaker over all energies compared to the limit shown in Fig. 8.6.

None of the limits calculated constrain so-called 'mid-range' cosmogenic neutrino fluxes, such as those predicted by [10].

8.2 Point Source Limits

The limits discussed in the previous section apply to the diffuse (i.e. total) flux of UHE neutrinos. Using the results of the ANITA-2 neutrino analysis, along with directionally dependent exposure of ANITA-2, it is possible to place limits on the luminosity of specific sources in UHE neutrinos.

Two candidate sources have been most commonly suggested as progenitors of UHE neutrinos: active galactic nuclei (AGN) and gamma-ray bursts (GRBs). A dedicated search for UHE neutrino emission from GRBs in the ANITA-2 data, that takes advantage of the short duration of GRB events to loosen thermal cut thresholds, has been produced [13]. Meanwhile, the tentative correlation of UHECR to AGN from Auger data [14] gives motivation to the calculation of UHE neutrino limits for AGN using the ANITA-2 results. Such limits are the focus in this section.

Most models of CR acceleration within dense sources, such as AGN, give rise to an associated neutrino flux via photo-meson production and subsequent meson decay (see e.g. [15] for a review). When considering the diffuse neutrino flux, no expectation is placed on the energy spectrum which is being constrained and the resulting flux limits are model-independent.[1] In the case of direct source limits, we consider sources accelerating charged particles via first order Fermi acceleration, providing a UHECR energy spectrum of $E_{CR} \propto E^{-2}$. The resultant neutrinos produced within the source will follow a similar energy spectrum. However, they will carry only a fraction of the energy of the primary CRs, leading to a lower energy cutoff in their $E_\nu \propto E^{-2}$ spectra [16].

With a model constraint on the expected source spectra, we can define the differential neutrino flux:

$$\frac{d\Phi_\nu}{dE_\nu} = \Phi_{90} \left(\frac{E_\nu}{E_0} \right)^{-\gamma} \tag{8.2.1}$$

Here, Φ_{90} is the spectrum normalisation on which we can place limits and has units $\mathrm{eV}^{-1}\,\mathrm{cm}^{-2}\,\mathrm{s}^{-1}$, while γ is the spectral index (2 for Fermi acceleration).

For a given region of the sky, we place limits on the flux at the 90 % C.L. according to a Poisson distribution, with the number of events we can exclude for the ANITA-2 flight being defined by:

$$N_{90} = \int_{E_{\nu,\min}}^{E_{\nu,\max}} \frac{d\Phi_\nu}{dE_\nu} A_{\mathrm{eff}} T_{\mathrm{live}} dE_\nu \tag{8.2.2}$$

Here, the combined effective area and livetime (A_{eff} and T_{live}, where A_{eff} accounts for the analysis efficiency) produce ANITA-2's exposure, which is a function of energy (see Fig. 8.7 for an example). We can therefore derive a constraint on the differential flux as:

[1] Constraints on the diffuse neutrino flux do assume that the energy spectrum is continuous within the energy interval being considered.

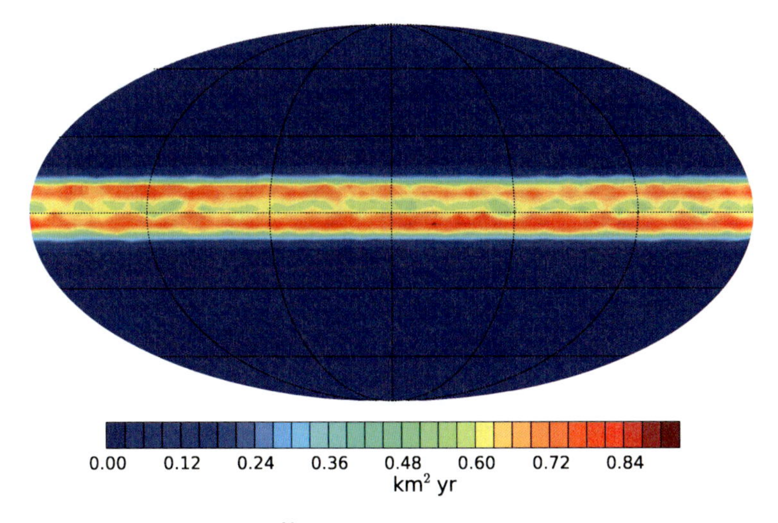

Fig. 8.7 ANITA-2's exposure at 10^{20} eV, using data from B. Mercurio and the *icemc* simulation

$$\Phi_{90} = \frac{N_{90}}{\int_{E_{\nu,\min}}^{E_{\nu,\max}} \left(\frac{E_\nu}{E_0}\right)^{-2} A_{\text{eff}} T_{\text{live}} dE_\nu} \tag{8.2.3}$$

Using Φ_{90}, it is then possible to set a limit on the flux and luminosity (assuming isotropic emission) using:

$$\Phi_{\nu,lim} = \int_{E_{\nu,\min}}^{E_{\nu,\max}} \Phi_{90} \left(\frac{E_\nu}{E_0}\right)^{-1} dE_\nu \tag{8.2.4}$$

$$L_{\nu,lim} = 4\pi d_s^2 \Phi_{\nu,lim} \tag{8.2.5}$$

where Φ_ν is the energy flux in eV cm^{-2} s^{-1}, d_s is the distance to source and L_ν is the source luminosity in eV s^{-1}.

8.2.1 Reflected Neutrino Search

Figure 8.8 demonstrates that the ANITA-2 experiment was optimally sensitive in the declination (δ) band $-13^\circ < \delta < 15^\circ$. However, the only currently published neutrino point source limits for AGN in the $E_\nu > 10^{19}$ eV regime are for Centaurus A (a nearby AGN) and Sagittarius A* (the Galactic centre) [17]. Both of these sources are outside of ANITA-2's optimal declination band, with $\delta < -13^\circ$.

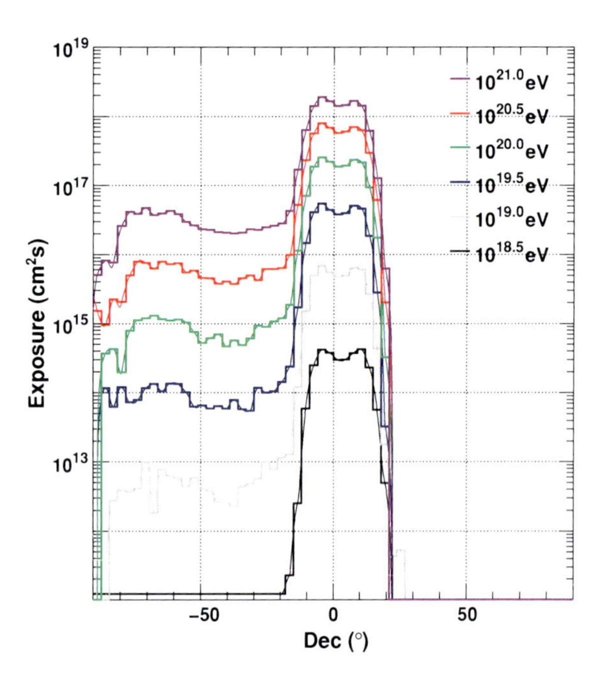

Fig. 8.8 ANITA-2's exposure as a function of declination (averaged over all right ascensions) for $10^{18.5} - 10^{21.0}$ eV

However, ANITA-2 was still sensitive to this region via reflected RF from down-going neutrinos.

The ANITA-2 analysis described in Chap. 6 contained a cut on elevation of $\theta > -35\,^\circ$. This cut was intended to remove events to which the antenna response was degraded. The elevation cut also reduced sensitivity to neutrino events viewed in reflection, particularly reflected signals from highly down-going neutrinos that would provide most of ANITA-2's sensitivity to sources with $\delta < -13\,^\circ$.

In order to place point source flux limits on sources that ANITA-2 was only sensitive to via reflected RF, further analysis was run with events passing all thermal analysis other than the elevation cut. All other thermal and clustering cuts remained unchanged. A summary of the downward-pointing ($\theta < -35\,^\circ$) events passing cuts and the results of event clustering is shown in Table 8.1.

A total of 66268 events passed thermal cuts with an elevation $< -35\,^\circ$, with the sample composed of 58 % VPOL dominated and 42 % HPOL dominated events. Event clustering results, shown in Fig. 8.9, returned zero isolated HPOL events and one isolated VPOL event, # 14250373. Inspecting this event, it is clear that unfiltered weak CW is present at 400 MHz in the direction that the event reconstructed. After filtering around the CW contaminated band in the range 380 MHz $< f <$ 420 MHz, the event reconstructs in a different direction and would not pass thermal cuts. Figure 8.10 shows that narrowband noise that was not filtered, along with interferometric images before and after filtering.

Table 8.1 A summary of the number of events in the downward-pointing sample and the outcomes of event clustering

	HPOL		VPOL	
	# cut	# remaining	# cut	# remaining
All down-pointing events	–	27857	–	38411
Clustered to flight/traverse	0	27857	0	27857
Clustered to base	27854	3	38404	7
Clustered to main analysis event	1	2	3	4
Clustered to down-pointing event	2	0	3	1
Remaining isolated events	–	0	–	1

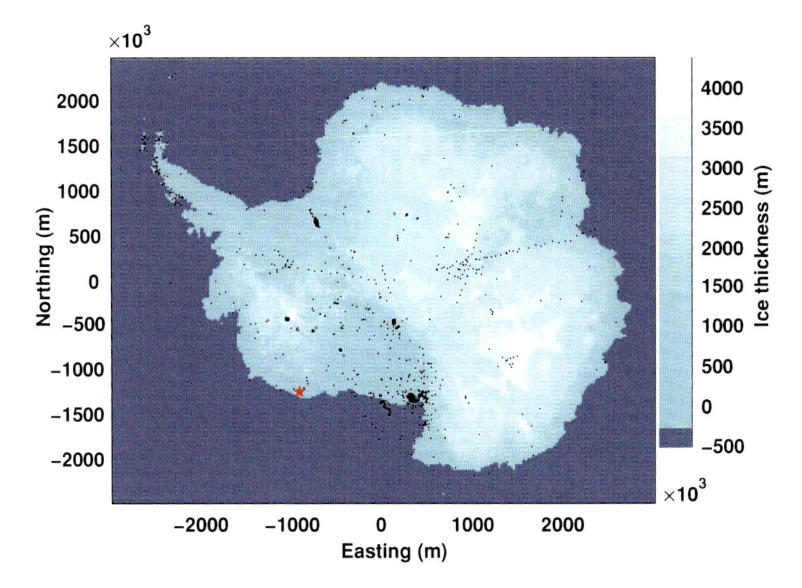

Fig. 8.9 Locations of the 66268 events with elevations $< -35°$ that pass all other analysis cuts. *Black* markers are events that cluster to bases, *green* are those that cluster to other events, the *red* marker near easting:northing $(-1000, -1250)$ is the single isolated event, #14250373

8.2.2 Neutrino Event Limits

In Sect. 6.5.1, the ANITA-2 pointing resolution for reconstruction of RF signals was demonstrated to be sub-degree in both azimuth and elevation for most correlation coefficient values. To calculate the arrival direction of a neutrino candidate, the RF signal, observed by ANITA-2 and pointed to ground via the analysis code, must be passed through the firn–air boundary with the refraction effect accounted for. Additionally, a polarisation measurement is required to decide which region of the Cherenkov cone was viewed by the instrument. The latter of these has a large associated uncertainty, resulting in poorer resolution on neutrino candidate pointing.

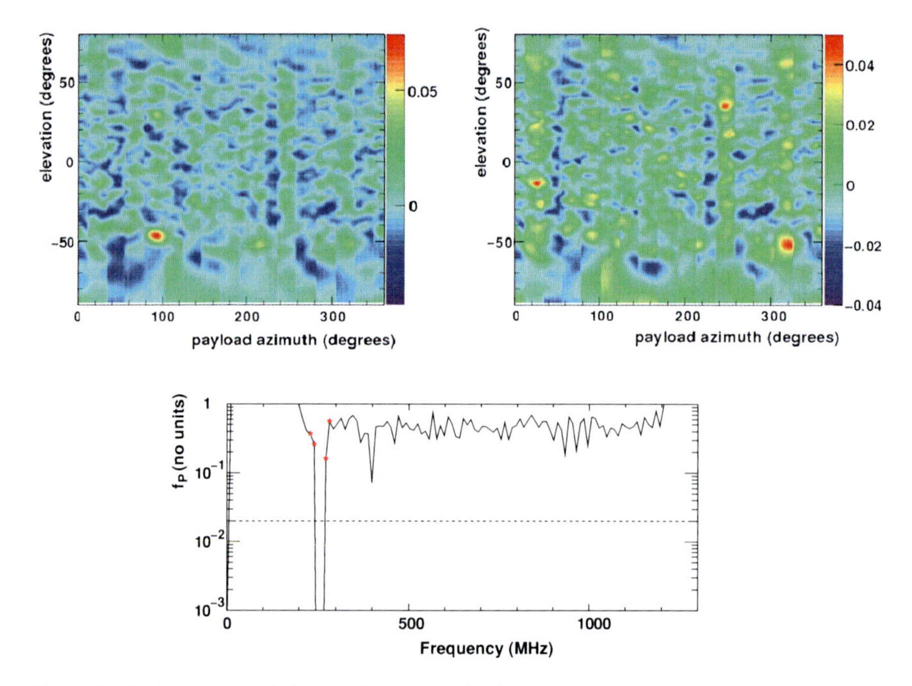

Fig. 8.10 *Top* interferometric images for event 14250373, with the event after normal analysis (*left*) and after filtering in a band 380 – 420 MHz. *Bottom* the ϕ-sector based filtering in the direction the event was recorded, the *dashed line* shows the analysis filter threshold, *red stars* indicate filtered bands

Due to the lack of any real data, the neutrino pointing resolution was calculated using simulated neutrinos, passed through the ANITA-2 analysis code. While the declination resolution is close to a degree, the right ascension resolution has suffered from uncertainty in the polarisation measurement, with an average resolution of about 7.7 °. The pointing resolution will clearly be much better for strong, well reconstructed events. As with the RF signal pointing resolution, neutrino pointing resolution is binned by peak correlation coefficient, results are displayed in Fig. 8.11.

Using the two VPOL events found in the ANITA-2 neutrino analysis, event limits can be set for all observable celestial coordinates. The events, # 15636066 and # 16014510, had P_{l} coefficients of ~0.25 and ~0.15 respectively. These P_{l} values have associated neutrino pointing resolutions of $\sigma_{RA} = 8\,°$, $\sigma_{Dec} = 1.5\,°$ and $\sigma_{RA} = 11\,°$, $\sigma_{Dec} = 1.5\,°$ respectively. These pointing resolutions can be used to set 90 % exclusion limits (N_{90}) on the number of events in Celestial coordinates, as shown in Fig. 8.12.

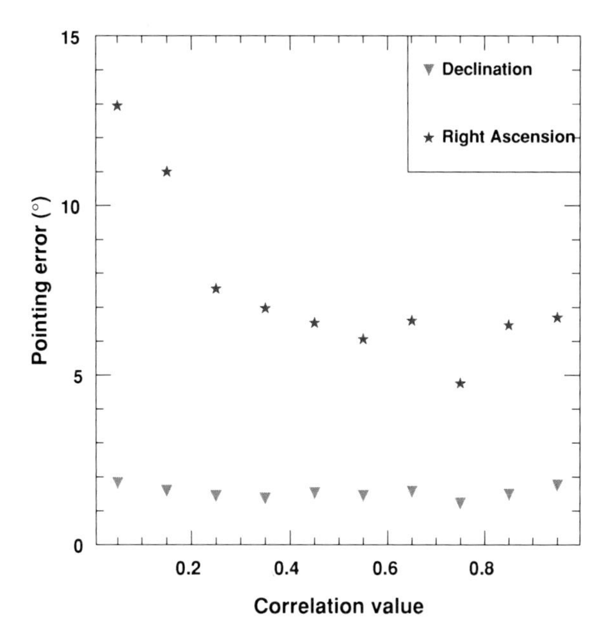

Fig. 8.11 The resolution on neutrino pointing as a function of peak correlation coefficient from the interferometric image ($P_{\rm I}$ value). Resolution for each angle is calculated using the root-mean-square, given in Eq. 6.5.1

8.2.3 Selected Source Limits

Using the 90 % C.L. of the number of events the analysis can exclude as a function of right ascension and declination, combined with the exposure of ANITA-2, limits can be placed on the neutrino flux for specific celestial coordinates.

UHE neutrino flux limits are places on AGN sources which lie within, or close to, ANITA-2's optimal sensitivity band. AGN are selected if they have been observed in γ-rays by the Fermi-LAT telescope [19] and are within a redshift of $z < 0.1$. Additionally, flux limits are calculated for Cen. A and Sgr. A*, as these sources are the only AGN for which UHE neutrino limits have previously been published [17]. Results for these sources are summarised in Table 8.2.

The Virgo cluster, the closest large cluster to the Milky Way, falls almost entirely in the optimal $-13° < \delta < 15°$ band, flux limits are also set for any AGN within this cluster. M87 is of particular interest as the central and most massive AGN in the Virgo cluster, it is also one of the AGN associated with a Fermi-LAT γ-ray observation [19]. Results for these sources are summarised in Table 8.3.

The model-dependent flux limits summarised in Tables 8.2 and 8.3 can be directly compared to limits from other astrophysical neutrino experiments. As mentioned previously, point source neutrino flux limits in the UHE regime have only been set for Cen. A and Sgr. A* [17]. Although published by members of the LUNASKA

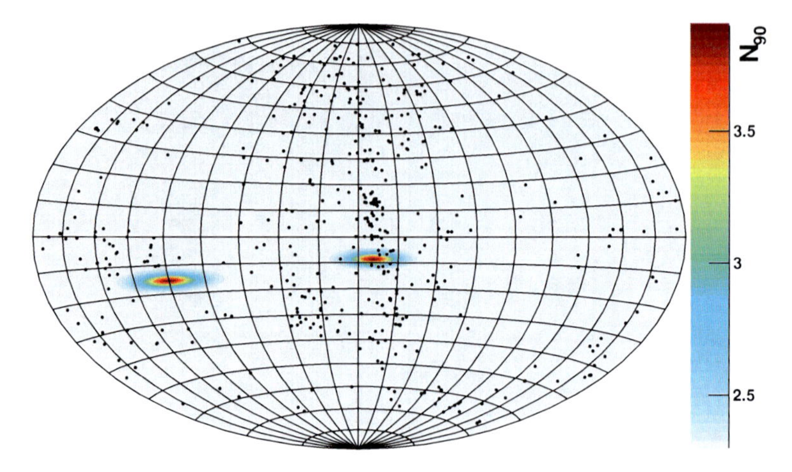

Fig. 8.12 The 90 % C.L. of the number of events ANITA-2 can exclude as a function of right ascension and declination. *Black dots* represent AGN within 100 Mpc, from the Veron-Cetty catalogue [18]

Table 8.2 UHE neutrino flux limits, in the form $E_{\nu}^{2}d\Phi/dE \leq \Phi_{90}(E/eV)^{-2}$, from the ANITA-2 experiment on AGN observed in γ-rays using the Fermi-LAT telescope and on Cen. A and Sgr. A*

Object	r.a. (°)	dec. (°)	z	Φ_{90}
1FGL J0339.1-1734	54.81	−17.6	0.07	30528
1FGL J1253.7+0326	193.45	3.44	0.07	327
1FGL J0308.3+0403	47.11	4.11	0.03	318.6
1FGL J2204.6+0442	331.07	4.67	0.03	311.6
1FGL J1551.7+0851	237.92	8.87	0.07	256.7
1FGL J1641.0+1143	250.25	11.73	0.08	364.2
1FGL J1230.8+1223	187.71	12.21	0.004	418.5
1FGL J0008.3+1452	2.02	14.84	0.05	1157.6
1FGL J1744.2+1934	265.99	19.59	0.08	12757
Cen. A	201.4	−43.0	0.0006	65714
Sgr. A*	266.4	−29.0	–	54580

collaboration, flux limits in [17] were compared (in some cases calculated for the first time) over a wide energy range and for a number of experiments. Flux limits were shown for three lunar Cherenkov experiments, LUNASKA, NuMoon [20] and GLUE [21], all of which had optimal sensitivities at $E_{\nu} > 10^{21}$ eV. Further limits were shown for IceCube [22] and Auger [23]. Finally, a limit was calculated for the radio Cherenkov experiment RICE [3] over a large $10^{16.5}$ eV $< E_{\nu} < 10^{22}$ eV range. For this work, the best limits for each energy range (IceCube, RICE and LUNASKA) are taken and compared with the point source flux limits from ANITA-2, with IceCube limits updated using more recent results [24]. The model-dependent and model-independent limits from ANITA-2 for the Cen. A and Sgr. A*, shown in

Table 8.3 UHE neutrino model-dependent flux limits, in the form $E_\nu^2 d\Phi/dE \leq \Phi_{90}(E/eV)^{-2}$, from the ANITA-2 experiment on AGN in the Virgo cluster

Object	r.a. (°)	dec. (°)	z	Φ_{90}
M 100	185.73	15.26	0.0052	1473
NGC 4383	186.36	16.27	0.0057	2969.3
NGC 4477	187.51	13.22	0.0045	592.2
M 84	186.27	12.2	0.0034	418.0
M 87	187.71	12.21	0.0042	418.5
NGC 4380	186.34	10.17	0.0032	275.6
M 49	187.44	8.13	0.0033	256.1
NGC 4651	190.93	16.28	0.0027	2983.8
M 58	189.43	11.18	0.005	320.0
NGC 4639	190.72	13.22	0.0033	593.1
IC 3576	189.16	6.1	0.0036	282.9

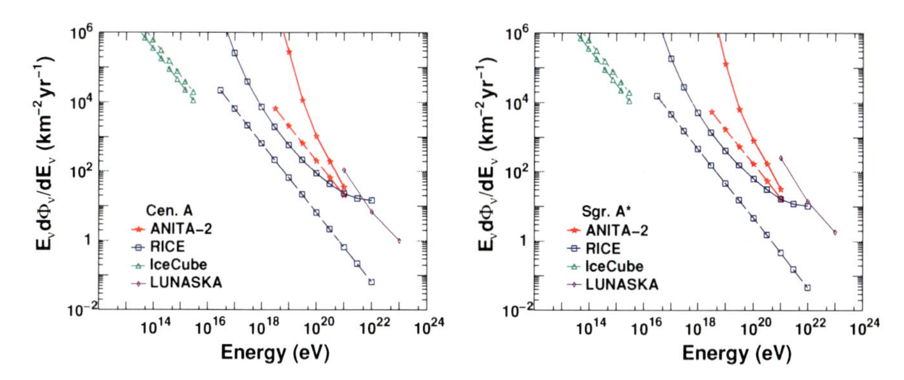

Fig. 8.13 Model-independent (*solid*) and -dependent (*dashed*) limits for Centaurus A (*left*) and Sagittarius A* (*right*) from ANITA-2, IceCube [24], LUNASKA [17] and RICE [3]

Fig. 8.13 are weaker than those from the RICE experiment, due to ANITA-2 viewing both sources in reflection.

The point source flux limit comparison is extended into ANITA-2's optimal declination range. M87 is chosen as a source for comparison due to it's relative proximity and the fact that it is the central galaxy of the nearest large cluster, Virgo (although it should be noted that Auger has seen no correlation of UHECRs to M87). Flux limits from ANITA-2 are compared to IceCube, for which model-dependent limits have been published for ν_μ and $\nu_\mu + \nu_\tau$ [24], and RICE, for which the elevation dependent exposure was calculated with data from [3] (D. Besson, Personal Communication). As ANITA-2 is the most sensitive UHE neutrino experiment to date, and M87 falls in the optimal declination band, both model-independent and model-dependent UHE neutrino flux limits from ANITA-2 for M87 are an order of magnitude more stringent than those from RICE. Limits for both experiments, along with those from IceCube, are shown in Fig. 8.14.

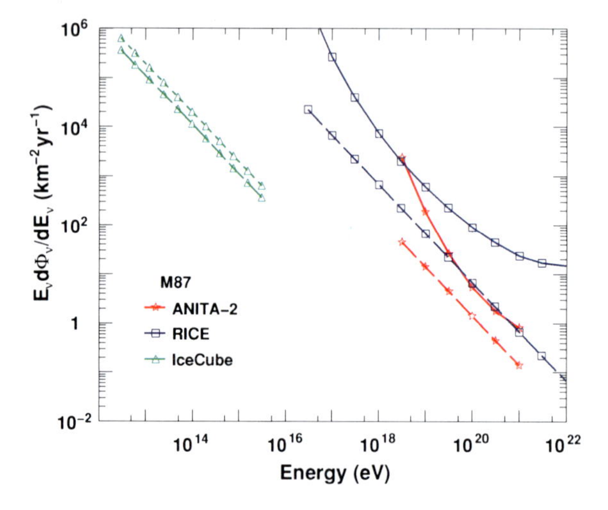

Fig. 8.14 Model-independent (*solid*) and -dependent (*dashed*) limits for M87 from ANITA-2, IceCube [24] and RICE [3]

8.3 Discussion

ANITA-2 is able to provide the most stringent limits to date on the UHE neutrino flux. A limit from the analysis outlined in [12] was more stringent than the limit from this analysis, due to the two isolated VPOL signals compared to one isolated VPOL signal in [12] and the fact that this analysis accounts for SURF saturation and includes more events at the anthropogenic cuts stage. The limit placed in this analysis is still more stringent than those from other experiments in the energy interval $10^{19.5}\,\text{eV} < E_\nu < 10^{22}\,\text{eV}$.

The point source flux limits set by the ANITA-2 experiment in this chapter are the first neutrino limits at $E_\nu > 10^{19}\,\text{eV}$ for AGN in the declination band $|\delta| < 20°$. It is within this band that ANITA-2 was most sensitive to neutrino induced Askaryan signals, with the limits calculated representing the most stringent from any UHE neutrino experiment to date.

The point source UHE neutrino limit results can be interpreted by comparing them to the UHECR flux from Auger [25], shown in Fig. 3.4. Given that ANITA-2 is sensitive to neutrinos with $E_\nu > 10^{18}\,\text{eV}$, and the neutrino produced from a cosmic-ray interaction is expected to have an energy of $E_\nu \sim 0.2E_{CR}$, the neutrino flux limits from ANITA-2 are compared to the cosmic-ray flux with $E_{CR} > 10^{19}\,\text{eV}$.

Consider a scenario in which the entire UHECR flux observed by Auger originates from M87 (or from the Virgo cluster as a whole). The implied luminosity of M87 in UHECR would be $L_{CR} = 2 \times 10^{42}\text{erg/s}$ for isotropic emission. The neutrino flux limits placed on sources in the M87 cluster imply an isotropic luminosity in neutrinos of $L_\nu < 1.5 \times 10^{44}\,\text{erg/s}$. In this single UHECR source scenario, the escape fraction of UHECRs from their source locations would be at least 1 %. This result in itself is not constraining. It demonstrates that, by comparing UHECR fluxes (along with a possible future positive correlation of UHECR arrival directions with specific sources) to neutrino flux limits, constraints on source optical depth could be calculated. This could then assist in enlightening physicists on specific acceleration locations and mechanisms within the UHECR sources themselves.

References

1. A. Connolly, R.S. Thorne, D. Waters, hep-ph/1102.0691
2. N.G. Lehtinen, P.W. Gorham, A.R. Jacobson, R.A. Roussel-Dupré, Phys. Rev. D **69**, 013008 (2004)
3. The RICE collaboration, I. Kravchenko et al., Phys. Rev. D **73**, 082002 (2006)
4. L.A. Anchordoqui, J.L. Feng, H. Goldberg, A.D. Shapere, Phys. Rev. D **66**, 103002 (2002), [hep-ph/0207139]
5. The ANITA collaboration, P. Gorham et al., astro-ph/1011.5004
6. G.J. Feldman, R.D. Cousins, Phys. Rev. D **57**, 3873 (1998), [physics/9711021]
7. The Pierre Auger collaboration, J. Abraham et al., Phys. Rev. D **79**, 102001 (2009)
8. The high resolution FlyGs eye collaboration, R.U. Abbasi et al., Astrophys. J. **684**, 790 (2008), [astro-ph/0803.0554]
9. The IceCube collaboration, R. Abbasi et al., Phys. Rev. D **83**, 092003 (2011), [astro-ph/1103.4250]
10. R. Engel, D. Seckel, T. Stanev, Phys. Rev. D **64**, 093010 (2001), [astro-ph/0101216]
11. O.E. Kalashev, V.A. Kuzmin, D.V. Semikoz, G. Sigl, Phys. Rev. D **66**, 063004 (2002), [hep-ph/0205050]
12. The ANITA collaboration, P.W. Gorham et al., Phys. Rev. D **82**, 022004 (2010), [astro-ph/1003.2961]
13. A. G. Vieregg et al., astro-ph/1102.3206
14. The Pierre Auger collaboration, P. Abreu et al., Astropart. Phys. **34**, 314 (2010), [astro-ph/1009.1855]
15. F. Halzen, D. Hooper, Rept. Prog. Phys. **65**, 1025 (2002), [astro-ph/0204527]
16. J.K. Becker, J. Phys. Conf. Ser. **136**, 022055 (2008), [astro-ph/0811.0696]
17. C.W. James et al., Mon. Not. R. Astron. Soc. **410**, 885 (2011), [astro-ph/0906.3766]
18. M. Véron-Cetty, P. Véron, Astron. & Astrophys. **518**, A10 (2010)
19. A.A. Abdo et al., Astrophys. J. **715**, 429 (2010), [astro-ph/1002.0150]
20. O. Scholten et al., Nucl. Instrum. Meth. **A604**, S102 (2009)
21. P.W. Gorham et al., Phys. Rev. Lett. **93**, 041101 (2004)
22. The IceCube collaboration, R. Abbasi et al., Phys. Rev. Lett. **103**, 221102 (2009), [astro-ph/0911.2338]
23. The Pierre Auger collaboration, J. Abraham et al., astro-ph/0906.2347
24. The IceCube collaboration, R. Abbasi et al., astro-ph/1012.2137
25. The Pierre Auger collaboration, J. Abraham et al., Phys. Lett. B **685**, 239 (2010), [astro-ph/1002.1975]

Chapter 9
Conclusions

A search for evidence of ultra-high energy neutrino and cosmic-ray interactions in the ANITA-2 data has been conducted. No statistically convincing evidence was observed of Askaryan emission from neutrino-induced particle cascades in the Antarctic ice sheet. After all analysis cuts two vertically polarised events consistent with the expected signals from neutrinos were discovered, on an expected background of 1.13 ± 0.27 events. The two events are therefore consistent with originating from the experimental background which is caused by either thermal or anthropogenic noise. In this instance, both events were separated by less than 50 km from the next nearest event that passed thermal cuts. Both events reconstruct to the Ross Ice Shelf, a region in which much of the human activity in Antarctica is located.

Using simulations of neutrino interactions and the ANITA-2 experiment, a limit on the neutrino flux was calculated that excludes certain models of cosmogenic neutrino flux at the 90 % confidence level. The limit placed demonstrates that ANITA-2 is the most sensitive experiment to neutrinos in the interval $10^{19}\,\text{eV} \leq E_\nu \leq 10^{22}\,\text{eV}$ to date. However, the limit placed on the diffuse UHE neutrino flux in this thesis is not as stringent as the published limit from previous analysis, outlined in [1, 2]. This is due to a combination of factors: two events were discovered in this analysis compared to one in the published analysis; more events were included in event clustering for anthropogenic noise rejection in this analysis; extra consideration of factors that would cause neutrino signals to be rejected were included in this analysis. Limits on the neutrino flux from specific active galactic nuclei were calculated. These represent the first neutrino limits on such sources with declinations of $|\delta| \leq -20^\circ$ and $E_\nu \geq 10^{19}\,\text{eV}$.

Although no evidence for neutrino interactions was detected, emission from geosynchrotron processes within cosmic-ray-induced air-showers was observed. After all analysis cuts four cosmic-ray events were observed on an expected background of 0.34 ± 0.15 events. The identity of the events was confirmed through the comparison of the measured polarisation of emission to the expected polarisation that would arise from geosynchrotron radiation. Additionally, cross-correlation of the cosmic-ray candidates' waveforms with one another demonstrated the similarity

M. J. Mottram, *A Search for Ultra-High Energy Neutrinos and Cosmic-Rays with ANITA-2*, Springer Theses, DOI: 10.1007/978-3-642-30032-5_9,

between the waveform shape of the cosmic-ray candidates, while the vast majority of other events passing thermal cuts in the horizontal polarisation did not correlate well with the cosmic-ray candidate waveforms. A further, non-isolated, event was also shown to be consistent with the expectation from air-shower geosynchrotron radiation.

The four isolated cosmic-ray events, and one further non-isolated candidate, extend the sample of 16 events observed by ANITA-1. ANITA-1 was the first experiment to measure cosmic-ray-induced geosynchrotron radiation above 600 MHz. A further flight of the ANITA experiment, with a dedicated cosmic-ray trigger, should be able to increase the sample of events by an order of magnitude. Such a dataset would compliment measurements by the Auger and HiRes experiments, with ANITA providing a new and complimentary detection method with expected sensitivity to UHECRs with $E > 10^{20}$ eV.

A third flight of the ANITA instrument will provide an opportunity to further constrain the cosmogenic neutrino flux. However, it appears likely that making actual observations of these particles will fall to successor experiments. One such experiment is the Askaryan Radio Array (ARA), an in-ice array of radio antennas. ARA test stations are being deployed at the South Pole, with much of the technology used deriving from ANITA.

References

1. The ANITA collaboration, P. W. Gorham et al., Phys. Rev. D **82**, 022004 (2010), [astro-ph/1003.2961]
2. The ANITA collaboration, P. Gorham et al., astro-ph/1011.5004

About the Author

Matthew Mottram received his first degree, MPhys Physics with Astrophysics, from the University of Leeds in 2003. He received his PhD for University College London for his work on the ANITA experiment. Matthew is currently working as a Post-doctoral research associate at the University of Sussex, working on the SNO+ experiment.

M. J. Mottram, *A Search for Ultra-High Energy Neutrinos and Cosmic-Rays with ANITA-2*, Springer Theses, DOI: 10.1007/978-3-642-30032-5,

Printed by Publishers' Graphics LLC
MO20120724